Ethics and Integrity in Educational Contexts

Volume 9

The aim of this series is to provide an authoritative series of books on topics relating to ethics and integrity in educational contexts. Its scope includes ethics and integrity, defined in broad and inclusive terms, in educational contexts. It focuses on higher education, but also welcomes contributions that address ethics and integrity in primary and secondary education, non-formal educational contexts, professional education, etc. We welcome books that address traditional academic integrity topics such as plagiarism, exam cheating, and collusion.

In addition, we are particularly interested in topics that extend beyond questions of student conduct, such as

- Quality assurance in education;
- Research ethics and integrity;
- Admissions fraud;
- Fake and fraudulent credentials;
- Publication ethics;
- Educational technology ethics (e.g., surveillance tech, machine learning, and artificial intelligence, as they are used in education);
- Biomedical ethics in educational contexts;
- Ethics in varsity and school sports.

This series extends beyond traditional and narrow concepts of academic integrity to broader interpretations of applied ethics in education, including corruption and ethical questions relating to instruction, assessment, and educational leadership. It also seeks to promote social justice, diversity, equity, and inclusion.

The series provides a forum to address emerging, urgent, and even provocative topics related to ethics and integrity at all levels of education, from a variety of disciplinary and geographical perspectives.

Irene Glendinning • Sarah Elaine Eaton
Beatriz Antonieta Moya Figueroa
Shiva Das Sivasubramaniam
Editors

Ethics and Integrity in Education (Research)

Derived from the 9th European Conference on Ethics and Integrity in Academia

Springer

Editors
Irene Glendinning
Independent Researcher
Lutterworth, UK

Sarah Elaine Eaton
Werklund School of Education
University of Calgary
Calgary, Canada

Beatriz Antonieta Moya Figueroa
Instituto de Éticas Aplicadas
Pontificia Universidad Católica de Chile
Santiago, Chile

Shiva Das Sivasubramaniam
School of Life and Health Sciences
University of Roehampton
London, UK

ISSN 2731-779X ISSN 2731-7803 (electronic)
Ethics and Integrity in Educational Contexts
ISBN 978-3-031-98156-2 ISBN 978-3-031-98157-9 (eBook)
https://doi.org/10.1007/978-3-031-98157-9

This Springer imprint is published by the registered company Springer Nature Switzerland AG
The registered company address is: Gewerbestrasse 11, 6330 Cham, Switzerland

Preface (Research)

Over the last two decades, academic integrity has become established as an important field for educational research. The scope for new research topics is huge, given the dynamic and diverse nature of this field. The ten chapters making up this book epitomise this diversity, including research on novel ways of conducting research into integrity, various aspects of the current topical subject of artificial intelligence (AI), to legal approaches to addressing educational fraud. The scope of chapters cuts across education levels and roles, with one study in secondary schools, several research projects into undergraduate students, studies into graduate students, research ethics, with some chapters focusing on educating and supporting academic staff. There are chapters specifically looking at the impact of generative AI (GenAI) on higher education with recommendations on how to effectively and ethically engage with this "double-edged" technology, which offers tremendous opportunities for innovation in education while posing significant threats to academic integrity if misused. Most importantly, the book includes a comparative literature analysis-based discussion on the ineffectiveness of AI "detection tools", with a timely and practical roadmap for institutions navigating the complexities of academic integrity in a GenAI-driven educational landscape.

This book is one of two complementary volumes that emerged from the same academic conference, held in Derby in May 2023 (please see the Acknowledgements below for more information). All 24 chapters across the two books explore a range of topics relating to academic and research integrity. This volume brings together chapters largely focusing on research; the second volume presents chapters that are practice-based.

Although this publication is edited by a team from the European Network for Academic Integrity (ENAI), the book contributors are truly international, with contributions from Bulgaria, Canada, Czechia, Chile, Saudi Arabia, South Africa, Sweden, and United Kingdom. The many themes covered by this volume are summarised in Fig. 1.

We will next summarise the ten chapters of the research volume in turn.

Although the opening chapter is badged as a literature review, it is much more involved than the title implies. The author Zander Janse van Rensburg, who is based

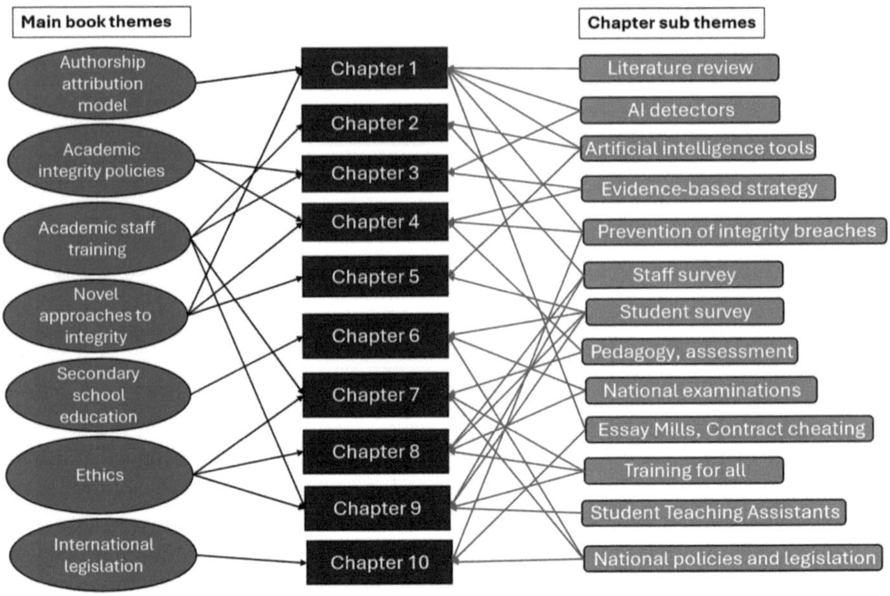

Fig. 1 Infographic summarising the main book themes and chapter subthemes

in North-West University, South Africa, analysed a range of published studies focused on evaluating the efficacy of artificial intelligence detection tools. It provides a comprehensive and timely analysis of AI detection tools, effectively integrating Actor-Network Theory (ANT) to illuminate the intricate relationships between students, educators, AI technologies, and institutional policies. The ANT approach does seem particularly suited to describe the progression stages in the adoption and normalisation of technological advances—currently at best we are barely beyond the earliest stage of *problematisation* considering the recent remarkable developments in GenAI. Its focus on the limitations of current AI detection systems, particularly when faced with obfuscation and adversarial techniques, offers valuable insights into the challenges of maintaining academic integrity in the age of GenAI. One of the chapter's strengths is its exploration of authorship attribution as a complementary technique to AI detection, which holds promise for capturing unique linguistic and stylistic traits of individual writers. The detailed literature review, supported by comparative analyses of existing detection tools and their efficacy metrics, is a significant contribution to understanding the complexities of AI in academic settings. Zander concludes with ideas on how authenticity of writing could be measured.

Chapter 2 is the first of two chapters concerning policies in Russell Group universities in England. The author, Caroline Campbell, discusses findings from a survey of academic staff that captures their perceptions of academic integrity and future requirements. The survey was run for two consecutive years, 2021–2022 and 2022–2023 in the University of Leeds. The findings provide valuable insights into the views of academic staff from across different subject areas, about students' knowledge and skills related to integrity, including perceptions of who should be responsible for educating students on these topics. Questions were added to the 2022–2023 run of the survey about GenAI, illuminating the broad variation in skills and experiences that respondents have for using AI within their teaching and research. The research confirms findings from other research about how much additional training is needed for everyone in education on both GenAI and academic integrity. The findings are thoughtfully analysed and have informed three actionable recommendations: (a) a comprehensive staff training program, (b) the development of targeted resources, and (c) a proactive institutional strategy. These measures aim to build a cohesive culture of academic integrity while addressing the challenges posed by the rapid evolution of GenAI. The emphasis on fostering shared understanding and collaborative practices across the higher education sector is particularly commendable. Overall, the chapter effectively bridges theory and practice.

The focus of Chap. 3 is also on provision of training. The authors Lauren Schrock, Asima Iqbal, and Alexa Kirkaldy are from University of Warwick, which, like Leeds University, is also an English Russell Group university. These authors present and evaluate the policies for academic integrity and the training needs of staff in their university. This chapter introduces an evidence-based approach to evaluating academic integrity training for educators by analysing their referrals of student assessments for potential misconduct. It highlights the importance of training to foster consistency in identifying and addressing academic misconduct while responding to emerging challenges such as GenAI. The work-in-progress methodology proposes collecting and analysing quantitative and qualitative data from referral databases to assess the impact of training on educators' decision-making. By aligning practices with institutional policies and refining training based on referral trends, the chapter offers a strategic framework to enhance academic integrity leadership and promote consistent, informed practices in higher education.

In Chap. 4 Jessica Kalra from British Columbia, Canada, proposes adapting the five levels of prevention used in medical settings for prevention of academic misconduct. The framing of these two very different contexts, together with the parallels drawn between them, provides potential for novel opportunities for developing preventative strategies. It advocates for a proactive, systemic approach to fostering academic integrity.

In comparison with preventive measures within medical sector, the chapter recommends (a) assessments prioritising learning over achievement, (b) cultivating a classroom culture of ethical scholarship through open communication, relationship-building, and raising awareness about responsible academic practices, (c) identifying the risks for misconduct and taking mitigation steps through tailored curricular & pedagogical interventions, (d) addressing misconduct with restorative practices, focusing on student support and reintegration rather than punishment, and (d) implementation of unbiased institutional policies with continuous reflective refinement. These interesting insights can be applied for helping and supporting individuals and institutions towards encouraging and promoting a culture of academic integrity.

Chapter 5 describes a new methodological approach to analysing qualitative data. The authors Michael Weiss, Sabiha Mumtaz, and Jamie Carmichael combine the use of Structural Topic Modelling (STM), ChatGPT, and human expertise to conduct the analysis. By demonstrating the advantages of STM and its application to real-world challenges in higher education, this chapter represents a forward-thinking approach to modern research. Readers will find it both insightful and practical, serving as a guide to unlocking the potential of qualitative and quantitative data in tandem. A module from the statistical tool R is used to conduct the STM. The method is used to analyse a combination of qualitative and quantitative data collected during a survey of postgraduate students in two campuses of a large international university. The authors recommend this more efficient approach as a way to encourage researchers to collect richer qualitative data.

Chapter 6 details a fascinating study of Swedish school pupils aged 14–16 from one school. The study was designed to explore the consequences of what the author Charlotta Rönn identifies as a contradiction or omission in the national school curricula in Sweden. According to the curricula, young learners in compulsory education are expected to become independent thinkers, in addition to working together effectively with other learners, and they must take responsibility for their learning. All work they submit needs to be assessed individually. However, there is no guidance for the teachers in how to assess work that has been co-written by two or more pupils. The study involved video recording of classroom activities across several subjects, followed by interviews with pupils and teachers. What the study revealed about student conduct is astonishing and unexpected. The findings will be especially interesting for people researching academic integrity. In addition, the findings are highly relevant to those countries relying on national examinations as a safe form of student assessment. In the analysis of the recordings, pupils demonstrated a range of sharing activities, of which the teachers were almost completely oblivious, including exchanging computers and "chained" communications for sharing answers to exam questions. Particularly shocking is the leaking of details about national tests and marking criteria, apparently as the result of teachers' carelessness. The leaked materials are routinely shared both within and across schools in Sweden, allowing pupils to pre-prepare their answers to specific questions. This makes a mockery of the national tests, which were designed to ensure secure and accurate summative assessments. Apart from the need to address the failings identified in nationally set educational requirements, the author raises important questions about how prepared this generation of young people will be for higher education, working life and ethical citizenship.

The two authors of Chap. 7, Irena Vassileva and Mariya Chankova from Bulgaria, present an evaluation of how policies for academic integrity and ethics in Bulgaria have advanced in the decade since the national report on Bulgaria was published for the IPPHEAE project. It critically examines academic ethics and integrity policies of 52 Bulgarian higher education institutions recognised by the National Evaluation and Accreditation Agency. The AIMM metrics used to in 2013 evaluate maturity on the national level, placed Bulgaria in 27th place out of the 27 countries included in this first study. It is clear that some progress has been made since then to address the deficits identified in policies. However, this new study demonstrates that institutions in Bulgaria have a lot more work to do in terms of developing and applying policies that are sustainable in the longer term and taken seriously by the academic communities they serve.

Such reforms are vital to ensuring a robust framework for academic integrity, which is pivotal for safeguarding institutional credibility and enhancing educational outcomes. This research provides a critical foundation for future efforts to standardise and improve academic ethics policies in Bulgaria, offering valuable insights for international stakeholders concerned with academic integrity in higher education.

A small study conducted by Shiva Sivasubramaniam, who has recently moved from the University of Derby to Roehampton University, UK, explores experiences of 35 researchers, from various countries, who are engaged in different roles, career stages, and research disciplines. The purpose of the research presented in Chap. 8 was to investigate compliance with ethical practices by exploring the conduct of the participants and their colleagues. Although the adherence to research conduct requirements appears to be generally good, the findings show a familiar pattern of questionable research practices (QRPs), bullying and academic misconduct, as observed in earlier similar studies. The chapter presents some relevant findings from the study to be used as guidance to improve support for graduate students, postdoctoral and early career researchers and to strengthen compliance with policies and principles relating to research conduct and ethical practices. These findings suggest that researchers involved in externally funded projects generally comply with ethical standards mandated by funding bodies and governmental organisations. The compliance with these standards is typically monitored through institutional audits. However, the study shows the need for institutions to address QRPs. The study concludes that institutions bear a critical responsibility to establish and enforce comprehensive policies to promote ethical research practices. Recommendations include enhancing procedural transparency, providing targeted training, and fostering a culture of integrity to minimise misconduct. This research underscores the need to adopt a proactive approach towards addressing ethical challenges. This will help to create robust and accountable research environments that support different disciplines.

In Chap. 9, the authors Tomáš Foltýnek and Martin Ukrop from Masarik University, Czechia, explore dilemmas faced by student teaching assistants (TAs) and measures taken to better prepare each new cohort of TAs for their forthcoming teaching experiences. The activities involved first conducting a questionnaire, using a range of scenarios to explore the views of teachers—including TAs—about potentially unethical situations. The analysis of results allowed the development of targeted materials, consisting of a teaching toolkit on ethical dilemmas, to be used in the training of all teachers. In addition, a Code of Ethics for Tutors was developed. The lessons learned and methods adopted for this study will be of interest to people in any country who employ students to support other students. The findings emphasise the need to incorporate ethics into teacher education programs and raise awareness of ethical issues across faculties. By proactively addressing these challenges, institutions can better support TAs in navigating their roles responsibly. The initiatives at Masaryk University offer a model for embedding ethics education into teaching programs, ensuring TAs are equipped to handle the unique dilemmas they face, ultimately fostering a more ethical teaching environment.

The tenth and final chapter in this book, by Michael Draper of Swansea University, Wales, presents the 2022 Recommendations of ETINED to counter education fraud. The chapter summarises the main points covered by the Recommendations and provides background about how they were created, validated, and verified and what factors influenced the contents of this document. The chapter specifically focuses on several more egregious types of education fraud, including contract cheating, and evaluates what is being done in some member states to counter these activities.

As can be seen from these brief summaries, this book includes a wealth of new ideas and research findings from around the globe. The topics include both practical and theoretical research, some small-scale local studies, and others, especially in the case of ETINED, much more extensive research, with far-reaching consequences. All the research presented can be applied in different settings and locations to help individuals and institutions to move forwards with integrity. The team of editors who put together this book hopes and expects that readers will find new ideas and great value in these chapters.

Lutterworth, UK — Irene Glendinning
Calgary, Canada — Sarah Elaine Eaton
Santiago, Chile — Beatriz Antonieta Moya Figueroa
London, UK — Shiva Das Sivasubramaniam

Acknowledgements: Ethics and Integrity in Education Proceeding Collection

As the lead and corresponding editor of the research and practice volumes of this book, I would like to express my heartfelt gratitude to all those who contributed to its completion. This book is the culmination of the engaging discussions, presentations, and workshops that took place during the 9th European Conference on Ethics and Integrity in Academia (ECEIA-2023), which I had the privilege of hosting at the University of Derby (UoD), UK in 2023. This project would not have been possible without the collaborative efforts of the authors, reviewers, and practitioners who shared their invaluable insights, research findings, and expertise.

Organised under the auspices of the European Network for Academic Integrity (ENAI), this conference highlighted critical issues at the intersection of emerging technologies and academic integrity. The editorial team extends heartfelt gratitude to Dr Tomáš Foltýnek, President of ENAI, and its Board members, as well as to Prof. Kathryn Mitchell, Vice-Chancellor of the University of Derby, and Prof. Chris Bussell, Pro Vice-Chancellor and Dean of College of Science and Engineering, for their invaluable support in making this conference and subsequent publication possible. The idea of transforming the conference proceedings into a book emerged from the importance of the themes explored during the event, particularly the impact of emerging technologies on academic integrity. When presented to Springer, the concept was met with enthusiastic encouragement, for which we are sincerely grateful to the editor Astrid Noordermeer. This support enabled us to approach contributors with a vision for the book, inviting them to expand their insights and arguments beyond their initial presentations. The response was overwhelmingly positive, leading us to produce not one but two volumes: one focused on research and the other on practical applications in academic integrity and ethics. We express our deepest appreciation to the authors for their contributions to these volumes. Their commitment to scholarly excellence has enriched this work immeasurably. We also thank the international experts and peer mentors who reviewed the chapters, offering constructive feedback and guidance that significantly enhanced the quality of submissions.

As lead editor, I was fortunate to collaborate with an exceptional editorial team: Dr Irene Glendinning, Prof. Sarah Elaine Eaton, Prof. Mary Davis, Dr Claudia

Baptista, and Dr Beatriz Antonieta Moya Figueroa, whose expertise and dedication were instrumental in shaping this book— I was merely "*standing on the shoulders of giants*". I extend my deepest gratitude to Dr Vadivel Parthsarathy for his invaluable support during the final stages of this project helping me with the submission of the research volume. His attention to detail and dedication during these last few days were essential to achieving the highest standards in the submission process.

This project embodies the spirit of scholarly collaboration across borders, featuring contributions from authors worldwide. Their diverse perspectives and expertise are reflected in the 24 chapters of both volumes, spanning critical areas of ethics and academic integrity. We hope that readers will find these volumes insightful and thought-provoking, furthering the global conversation on these vital topics.

Dr Shiva Das Sivasubramaniam

Corresponding Editor (On Behalf of the Editorial Team)

Contents

Contributors

Caroline Campbell University of Leeds, Leeds, UK

Jamie J. Carmichael Carleton University, Ottawa, ON, Canada

Mariya Chankova South-West University Neofit Rilski, Blagoevgrad, Bulgaria

Michael Draper Swansea University, Swansea, UK

Tomáš Foltýnek Faculty of Informatics, Masaryk University, Brno, Czechia

Asima Iqbal WMG, University of Warwick, Coventry, UK

Jessica Kalra Faculty of Pharmaceutical Sciences, University of British Columbia, Vancouver, BC, Canada

Alexa Kirkaldy WMG, University of Warwick, Coventry, UK

Sabiha Mumtaz University of Wollongong in Dubai, Dubai, UAE

Zander Janse van Rensburg North-West University, Potchefstroom, South Africa

Charlotta Rönn Linnaeus University, Växjö, Sweden

Lauren Schrock WMG, University of Warwick, Coventry, UK

Shiva Das Sivasubramaniam School of Life and Health Sciences, University of Roehampton, London, UK

Martin Ukrop Faculty of Informatics, Masaryk University, Brno, Czechia

Irena Vassileva New Bulgarian University, Sofia, Bulgaria

Michael Weiss Carleton University, Ottawa, ON, Canada

Chapter 1
A Literature Review on AI Detection: Investigating the Potential of Authorship Attribution Methods

Zander Janse van Rensburg

Abstract Plagiarism, cheating, and other forms of academic dishonesty have become commonplace, making academic misconduct an increasingly urgent problem to mitigate in higher education. Additionally, AI-assisted writing technologies will increase, significantly impacting sustainable strategies to adapt to such technological revolutions. This chapter analyses the efficacy of AI detection tools in publicly available software, using Actor-Network Theory (ANT) as a framework to explore the complex relationships between students, educators, AI technologies, and institutional policies. This chapter examines recent studies on AI detection with a special focus on their objectives, approaches, and sampling techniques used in recent studies on AI detection efficacy, primarily to investigate the possibilities of varied results among these studies. The results of these studies point to promising results where the content is generated purely by GenAI. However, some studies found that the performance of AI detection is significantly impacted by adversarial techniques and obfuscation. Considering this, this chapter proposes the implementation of authorship attribution as a complementary technique in AI detection. By focussing on unique linguistic and stylistic characteristics of individual writers, authorship attribution offers a more reliable method of identifying academic misconduct. Integrating AI detection with authorship attribution proposes a more effective way to detect AI-generated content and strengthen academic integrity. This chapter advocates for a comprehensive approach that combines advanced detection techniques with institutional policies to meet the evolving challenges of generative AI technologies.

Keywords AI detection tools · Authorship attribution · Academic integrity · Generative AI (GenAI) · Actor-network theory (ANT) · Higher education misconduct

---

Z. J. van Rensburg (✉)
North-West University, Potchefstroom, South Africa
e-mail: Zander.jansevanrensburg@nwu.ac.za

I. Glendinning et al. (eds.), *Ethics and Integrity in Education (Research)*, Ethics and Integrity in Educational Contexts 9,
https://doi.org/10.1007/978-3-031-98157-9_1

Introduction

Generative AI (GenAI) has given humanity endless possibilities to transcend traditional information processing capabilities, which have wide-ranging implications for higher education. On the one hand, GenAI's information processing and generation capabilities offer numerous advantages, such as personalised learning experiences (Pesovski et al., 2024), efficient content creation, enhanced research capabilities (Bolanos et al., 2024), etc. On the other hand, the ease of generating text with AI tools can lead to increased instances of plagiarism and cheating (Perkins et al., 2024); reliance on AI-generated content can potentially hinder students' development (Essien et al., 2024); and most relevant to this chapter, the current AI detection tools fail to accurately identify AI-generated content, leading to possible issues in maintaining academic integrity (Bellini et al., 2024). While GenAI challenges learning and authentic human-generated content, it simultaneously presents opportunities for innovation in educational practices, authentic learning, and integrity.

However, for higher education to maintain academic integrity has been primarily based on reactive measures, which have traditionally been heavily based on plagiarism detection software such as Turnitin (Mphahlele & Mckenna, 2019). This practice, while unfortunate, frequently identifies instances of plagiarism and cheating (i.e., text-lifting, patchwriting, etc.) after the fact. Proponents of this culture focus more on detection and penalty than on fostering an environment of proactive education and ethical scholarship (Hass, 2022). Higher education is now confronted with more sophisticated forms of plagiarism and cheating, as opposed to simpler lifting techniques. Therefore, with the advent of GenAI, the nature of misconduct has evolved, necessitating the adaptation of detection strategies.

To this end, the evaluation of AI detection methods has become crucial in upholding the integrity of academic work and ensuring the ethical use of GenAI within educational settings, serving as a measure to assess students' skills and knowledge acquisition. Consequently, researchers soon started to test available tools, unfortunately with varying results (Bellini et al., 2024; Elkhatat et al., 2023; Perkins et al., 2023b; Perkins et al., 2024; Popkov & Barrett, 2024; Weber-Wulff et al., 2023). As evidenced by this research, simple obfuscation or adversarial techniques could be used to trick AI detection models. An additional challenge is the ability to effectively detect AI influence after the occurrence of the complex processes by which academic work is compiled, which is by nature or should ideally be a recursive process (Eckstein et al., 2011). In other words, AI detection performs well when texts are lifted directly from these models, but when various stages of writing (planning, drafting, revising, and editing) occur, the text inevitably takes on more human influence, causing AI detection rates to drastically drop. Although planning, drafting, revising, and editing are integral to the learning process, these stages do not preclude plagiarism or cheating via GenAI. For instance, students can still generate large portions of their text using AI and superficially edit it, i.e. applying obfuscation or adversarial techniques. Here, the final product could be altered to appear more human while bypassing academic integrity checks. This underscores the need for higher education institutions to invest in continuous analysis, development, and refinement of AI detection strategies. However, more importantly, acceptable levels of GenAI influence still need to be defined in the higher education sector. This chapter adopts a framework that aims to analyse the complex interplay between various actors involved in AI detection. Actor-network theory (ANT) provides such a framework by focussing on the relations and networks between human and nonhuman entities. Primarily developed by Bruno Latour, ANT treats all entities, such as students, educators, GenAI, AI detection tools, and institutional policies, as actors within a network that continuously interact and influence each other (Latour, 2005).

This perspective is particularly relevant for examining the efficacy of AI detection software and its application because it highlights how these tools function within the broader context of educational practices and policies. In doing so, ANT allows us to map interactions and negotiations between different actors, revealing the strengths and limitations of current AI detection methods. It seems that AI detection tools might work as designed, but human interpretation and interventions after AI content generation contribute to their perceived failure. Therefore, we must adopt new approaches, moving away from purely reactive measures. Exploring avenues such as understanding the nuances of student writing and how AI influences their processes, particularly through authorship attribution, may provide more effective solutions.

Following this position, this chapter will perform a literature review of existing research on the efficacy of publicly available AI detection tools, within the theoretical framework of ANT. Hereby, providing insights into their current performance and potential areas for enhancement. Finally, this chapter will propose authorship attribution as a probable solution to the current AI detection strategies.

Actor-Network Theory as a Theoretical Framework for Situating AI Detection Technology in Academia

Serving as a tool for initial inquiry, ANT provides a lens for understanding the complexity of social, technological, and natural interactions by focussing on the relationships and networks formed between various entities, both human and nonhuman, in that it traces how 'human and nonhuman elements are enacted as they become assembled into collectives of activity' (Fenwick, 2010, p. 120). In a network, actors or actants are defined as objects or entities whose presence or absence can be felt independently by other actors (Tatnall & Burgess, 2002) and translate their will into the language of their interests (Páscoa & Tribolet, 2014). Simply put, a network in ANT can be seen as an association of heterogeneous elements (actors/actants) (Tatnall, 2009). By its nature, any given network remains dynamic and ever changing as the actors' resources, interests, and viewpoints are continuously changing and, as a result, their relationships strengthen or weaken within the network (Paledi & Alexander, 2017).

To create an actor network, the interests of the actors must be aligned through the four-step translation process: (1) *problematisation,* here, the problem is defined accompanied by a proposed solution; (2) *interessement,* efforts are made to secure actors' commitment to their roles; (3) *enrolment,* actors accept and adopt their roles; and (4) *mobilisation*, where the network stabilises, and proposed solutions becomes a 'black box', which refers to a system whose inner workings are opaque but accepted as functional (Keith & Van Belle, 2014). The primary actors within this network are students, educators, GenAI technologies, AI detection tools, and governance, as presented in Fig. 1.1. These actors interact and influence each other, shaping the efficacy and implementation of AI detection to maintain academic integrity. Identifying these actors sets the stage for analysing the translation process within the network.

Figure 1.1 distinguishes between *central* and *supporting actants* based on their proximity to the core processes of maintaining academic integrity. The *central actants*—students, educators, AI detection tools, academic integrity, and governance—are directly involved in shaping and negotiating outcomes. These actors interact frequently and have the most direct impact on the effectiveness of AI detection and the preservation of academic integrity.

On the contrary, the supporting actors, including policies, systems, training, and technology, provide the essential frameworks, resources, and mechanisms that enable the central actors to operate effectively. For example, governance refers to authority and decision-making structures, while systems represent the operational mechanisms through which governance is enacted. Similarly, policy sets the rules

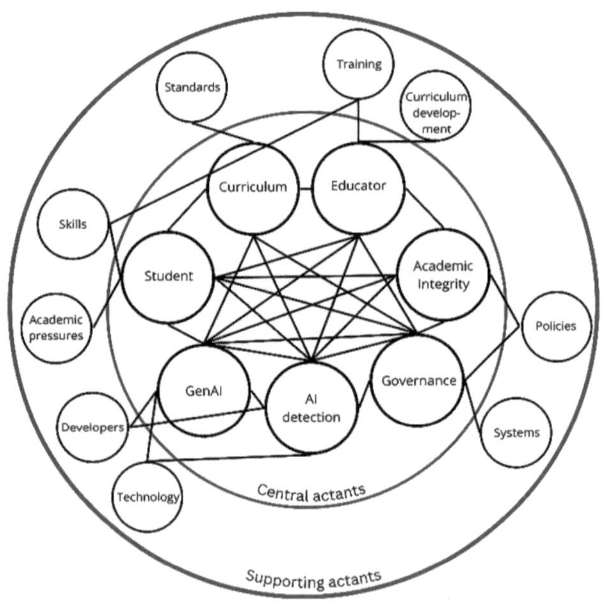

Fig. 1.1 Rudimentary AI-detection actor-network

that govern education, while curriculum refers to the educational content delivered to students. These supporting actants are critical to the functioning of the central network, but their influence is more indirect, shaping the broader environment in which the central actors engage. This layered structure clarifies the various levels of interaction within the actor network, with central actants driving the immediate outcomes and supporting actants creating the necessary conditions for effective operation.

What follows is an analysis of the translation process within this actor network, referring to how ideas and actions are negotiated among different actors. It will become evident that, concerning AI detection techniques, higher education has not yet fully reached the second stage, where interests across the network align and stabilise. First, problematisation involves defining the problem and positioning the focal actor as indispensable to solving it, establishing an Obligatory Passage Point (OPP) that all actors must pass through. For example, in this context, AI detection technologies could be positioned as essential to maintain academic integrity, making the adoption of these tools the OPP to ensure credible assessments. Second, *interessement* then refers to the strategies used to establish other actors into their roles, such as demonstrating the effectiveness and necessity of AI detection tools to educators and administrators. Third, *enrolment* follows, where these actors accept their roles and align their actions with the goals of the network, leading to widespread adoption of the technology. Finally, mobilisation occurs when the solution gains widespread acceptance and becomes a black box, meaning that it is taken for granted and is no longer questioned by the actors involved. This process ensures that AI detection tools are fully integrated into the academic environment, facilitating smoother and more effective monitoring of academic integrity.

Although the translation process has been outlined here, it is important to note that it is by no means complete due to the current shortcomings and expectations surrounding AI detection. Therefore, after a discussion of the efficacy of these tools, a more detailed discussion will follow, addressing probable reasons for their limitations and exploring potential improvements. This analysis will provide a deeper understanding of the challenges faced and guide future developments in AI detection technology.

Methodology

A review of the literature on AI detection methods was conducted for this chapter. The inclusion criteria for this review were peer-reviewed studies published after November 2022 when ChatGPT was available to the public. Additionally, a focus was placed on studies that tested AI detection methods with publicly available on-the-market tools. The primary reason for this approach was that these tools would likely be widely used in higher education and could have a significant impact on decision making in the sector. A comprehensive search was conducted across several databases, including Scopus and Web of Science. The keywords used were "AI detection and higher education." An initial search yielded numerous studies (Scopus: 23 documents; Web of Science: 3 documents) using the following search queries.

Scopus
("AI detection" OR "artificial intelligence detection") AND ("higher education" OR "university" OR "academic integrity" OR "plagiarism detection" OR "AI tools in education" OR "academic misconduct")
Web of Science:
("AI detection" OR "artificial intelligence detection") AND ("higher education" OR "university") AND ("academic integrity" OR "plagiarism detection" OR "AI tools in education" OR "academic misconduct")

These studies were then screened by titles and abstracts for relevance. Full-text articles were retrieved for those that met the inclusion criteria. Data were extracted from selected studies, focussing on the objectives, methodologies, key findings and implications of the study. Hereafter, a comparative analysis of the study design, results, discussion, and limitations was performed to provide a broader understanding of the possibility of AI detection.

Efficacy Assessments of AI Detection Tools

Given the importance of the impact of AI on higher education and the ongoing efforts to integrate AI into curricula, it is essential to take note of various efforts to measure their effectiveness. As evidenced by Perkins et al. (2023a) and Bekker (2024), higher education will stand firm in ensuring that human-centric learning remains. Consequently, robust assessment strategies to measure human learning and understanding will remain essential (Bearman et al., 2024; Hosseini et al., 2023). Consequently, AI detection tools will need to be continually tested and developed so that they would be able to distinguish between human and AI-generated content as AI outputs become more human-like. Researchers and developers may need to shift focus towards distinguishing the qualities of human writing in terms of the production process and the characteristics of the final product.

Overview of AI Detection Efficacy Studies: Objectives, Approaches, and Sample Techniques

This section aims to provide an overview of the objectives, approaches, and sampling techniques used in recent AI detection efficacy studies. Since the field of AI detection in academic work is still in its infancy, the following studies make foundational contributions that lay the groundwork for the refinement of AI detection methodologies. Initially, I will discuss binary classification approaches. Binary classification approaches provide a straightforward measure of AI detection accuracy, allowing a clear assessment of performance based on predetermined criteria such as sensitivity, specificity, error rates, and predictive value. Following this, I will discuss studies conducted by means of evaluating detection tools under conditions of obfuscation and adversarial techniques, which, in turn, offer insights into the robustness and adaptability of AI detection tools in real-world scenarios where deceptive practices and sophisticated manipulation techniques are prevalent. Although binary classification provides a foundational understanding of AI detection capabilities, evaluation under adversarial conditions helps identify potential weaknesses and informs the development of more resilient detection strategies. The aim of this section is to demonstrate the complexity of AI detection and point readers to probable approaches to enable them to replicate these approaches in their own environments since the product of academic work is highly context-specific. As a result, following this line of scholarly inquiry, it appears that stylistic and linguistic variance could be a fruitful approach in the refinement of AI detection.

Two key features in the AI detection tools analysed: (a) many of the tools use *binary classifications*—categorising text as either human or AI-generated—and (b) several studies in the list employ *obfuscation* and *adversarial techniques* to evaluate the tools' ability to detect manipulated content. These considerations are critical in understanding the limitations and accuracy trends that emerge from the research community's focus on specific tools. The first observation I would like to illustrate in Table 1.1 is a consolidated view of the frequency and distribution of AI detection tools as tested in recent efficacy studies, illustrating the research community's emphasis on specific tools and highlighting trends in their application across different academic inquiries.

Binary Classification Approaches for Assessing AI Detection Efficacy

The following studies contributed to the field of AI detection in disciplinary-specific contexts and initial evaluation metrics. Firstly, Elkhatat et al. (2023) conducted a study to investigate the capabilities of five AI content detection tools to differentiate between human-written and AI-generated content, specifically in terms of their specificity, sensitivity, positive predictive value and negative predictive value. Generated by ChatGPT-3.5 and ChatGPT-4, the authors generated 15 paragraphs using the following prompt: "write around 100 words on the application of cooling towers in the engineering process." Furthermore, they used five human-written samples as control samples, which were chosen from the introduction sections of five distinct lab reports written by undergraduate chemical engineering students submitted in 2018.

Second, Bellini et al. (2024) conducted their experiment in the field of anaesthesia and intensive care, whereby they aimed to test four tools in terms of sensitivity

Table 1.1 AI detection tools in recent efficacy studies

Frequency in studies	AI detection tools tests	Weber-Wulff et al. (2023)	Perkins et al. (2024)	Chaka (2023)	Elkhatat et al. (2023)	Bellini et al. (2024)	Popkav and Barrett (2024)	Perkins et al. (2023b)
5	GPTZero	x	x	x	x	x		
4	Writer.com	x		x	x	x		
4	ZeroGPT	x	x			x	x	
3	Turnitin	x	x					x
3	Copyleaks		x	x	x			
3	OpenAI classifier	x		x	x			
3	Crossplag	x	x		x			
2	Originality. AI					x	x	
2	GPT-2 output detector	x	x					
1	DetectGPT	x						
1	Content at scale	x						
1	Compilatio	x						
1	Go Winston	x						
1	Writeful GPT detector	x						
1	Giant language model test room			x				
1	Check for AI	x						
1	Plagiarism checker	x						
1	GPTKit		x					
Total:		14	7	5	4	4	2	1

in correctly identifying AI-generated text and specificity in correctly identifying human-generated text. Their sample was composed of three batches of text: (1) with ChatGPT-4 four short scientific articles; (2) again with ChatGPT-4 reformulating the first four texts with specific instructions to avoid being detected; and (3) one human author formulating two texts authored without AI intervention.

Third, Popkav and Barret's (2024) study aimed to assess the precision of AI detection tools to distinguish human-written content from AI-generated content within academic writing on behavioural health. It specifically evaluated the possibility of false positive rates on human-generated text and false negative rates on AI-generated text produced from free and paid AI detection tools. Their sample

Table 1.2 Evaluation classification for AI detection (Bellini et al., 2024; Elkhatat et al., 2023; Popkav & Barret, 2024)

Criteria	Metrics	Description
Performance	Sensitivity / true positive rate (TPR)	The proportion of AI-generated content correctly identified by the detectors out of all AI-generated content (Bellini et al., 2024; Elkhatat et al., 2023; Popkav & Barret, 2024)
	Specificity / true negative rate (TNR)	The proportion of human-generated content correctly identified by the detectors (Bellini et al., 2024; Elkhatat et al., 2023; Popkav & Barret, 2024)
Predictive value	Positive predictive value (PPV)	The proportion of positive results in statistics and diagnostic tests that are actually positive results (Elkhatat et al., 2023)
	Negative predictive value (NPV)	The proportion of negative results in statistics and diagnostic tests that are accurate negative results (Elkhatat et al., 2023)

included the following: (1) 100 research articles published in the field of behavioural health and psychiatry journals; and (2) 200 AI-generated samples consisting of 500–600 words, where 100 samples were generated by ChatGPT-4, and 100 samples from Claude. These AI tools were provided with the results sections of academic papers used for the human-written sample and instructed to generate discussion sections based on the results and in the academic writing style of the same human authors.

These three studies included controls of human-written texts, along with purely AI-generated text, although with instructions to avoid detection. However, what is of interest here are the metrics chosen by these researchers to evaluate the performance of these tools. In Table 1.2, the various binary classifications were grouped under two criteria, namely, performance rates and predictive value. Each with two metrics to measure specific rates or values, as described below.

These studies are some of the first studies to assess publicly available AI detection tools. While they used relatively small sample sizes in controlled environments, they provide crucial insights into the initial capabilities and limitations of these tools in how they can accurately identify the creators of content.

The approaches taken by these researchers, which are focussed on sensitivity, specificity, and predictive values, underscore the complexity of AI detection in distinguishing between human and AI-generated content.

Evaluating AI Detection under Conditions of Obfuscation and Adversarial Techniques

In light of the notion of the recursive nature of academic work or wilful misconduct, it is critical to try and simulate real-world academic writing scenarios when testing AI detection tools. Weber-Wulff et al. (2023) were one of the first groups to test the efficacy of 12 publicly and commercially available AI detection tools, especially by obfuscating the text generated by AI. Their objective was to test detection tools that differentiate between human-written text and ChatGPT-generated text and to see how machine translation and content obfuscation may affect detection accuracy. Their test sample comprised a total of 54 simulated test cases, divided among five categories: (1) 10,000 human-written characters on the level of an undergraduate across varying fields; (2) 10,000 human-written texts translated into English written in Bosnian, Czech, German, Latvian, Slovak, Spanish, and Swedish; (3) minimum of 2000 characters generated by ChatGPT; (4) generated a minimum of 2000 characters by ChatGPT 3.5, then manually edited as a means to simulate probable obfuscation methods students may apply; and (5) the aforementioned AI-generated was then machine paraphrased with Quillbot. Their sample variety was compiled to mimic potential cases in academic settings, such as machine translation between languages, manual editing, and paraphrasing tools to trick detectors.

Along the same lines of Weber-Wulff et al. (2023), Perkins et al. (2024) conducted a comprehensive analysis to evaluate the efficacy of AI detectors when they are subjected to adversarial techniques. Their sample consisted of three batches of sample text, 114 in total: (1) 15 initial samples generated by GPT4, Claude 2 and Bard, five each; whereby each model was prompted to cover a range of topics, linguistic styles, complexity levels, and genres; (2) ten control samples were human-generated by the researchers, without AI influence; and (3) the 15 initial samples were modified using six adversarial techniques, amounting to 89 samples due to technical constraints. The adversarial techniques, applied by GPT4, Claude 2, Bard, and Quillbot, included adding spelling errors (SE), increasing burstiness (IB), increasing complexity (IC) of writing styles, decreasing complexity (DC) of vocabulary and sentence structures, writing as a non-native English speaker (NNES), and paraphrasing (PR). Although Weber-Wulff et al. (2023) used only one model to generate texts, Perkins et al. (2024) used three models to allow for stylistic and linguistic variance between the models. In addition, Perkins et al. (2024) added more complexity to obfuscation or adversarial techniques; however, admittedly, all obfuscation techniques were applied by GenAI.

Table 1.3 represents a consolidated view of the criteria and metrics used by both research groups, demonstrating their commitment to moving beyond binary classification towards a more nuanced understanding of how AI detectors could be tested.

The layered complexity of obfuscation and adversarial techniques, which are measured by these researchers, poses a significant challenge to the interpretability of AI detection reports, since, as we will discuss in the upcoming sections, most detection systems follow a binary classification reporting format which does not encapsulate the complexity of the academic text. As a result, academic decision-making ability is hindered, in the case, where the recommendation of the detection tools is followed without further deliberation.

Table 1.3 Evaluation metrics for AI-detection (Perkins et al., 2024; Weber-Wulff et al., 2023)

Criteria	Metrics	Description
Overall accuracy	Binary accuracy (Acc_Bin)	Measures the ratio of texts correctly classified as human-written or AI-generated to the total number of cases, disregarding partially correct classifications, Positive (P) or Negative(N) (Perkins et al., 2024; Weber-Wulff et al., 2023)
	Binary inclusive accuracy (Acc_Bin_Incl)	Partially correct classifications (Partially True Positives and Partially True Negatives) as correct in the total accuracy calculation (Weber-Wulff et al., 2023)
	Semi-binary accuracy (Acc_SemiBin)	Considers partially correct classifications (Perkins et al., 2024; Weber-Wulff et al., 2023)
	Logarithmic accuracy (Acc_Logarithmic)	Assigns a weighted score to different levels of classification correctness, allowing for a more nuanced assessment of the tool's discriminative capabilities (Perkins et al., 2024; Weber-Wulff et al., 2023)

Detection in Multilingual Contexts

Exploring the accuracy of AI content detection against AI-generated multilingual text across various AI platforms underscores the need for detection capabilities in multicultural contexts. Furthermore, this could also shed light on the potential to detect AI-generated text subject to adversarial techniques similar to translation plagiarism (Tauginienė et al., 2018). To this end, Chaka (2023) employed an exploratory design by generating content using ChatGPT, YouChat, and Chatsonic in the context of English-language studies as a means of measuring the precision of five content detection tools. As a second objective, the researcher tested the efficacy of GPTZero and Copyleaks in detecting the translation of German, French, Spanish, Southern Sotho, and isiZulu by Google Translate. This researcher prompted the three GenAIs with seven distinct prompts, which resulted in 21 unique base samples, which would later be machine-translated into the languages mentioned above. Chaka found mixed results with all tools, GPTZero and Copyleaks performed best. Interestingly, both GPTZero and Copyleaks demonstrated moderate success with German, French, and Spanish. Whereas, they failed with the African languages, which is expected, since these languages are not widely supported. Chaka's approach is significant because it takes on the challenge of how machine translation can obfuscate AI writing patterns and tests the robustness of detection tools in a multilingual setting. It is, therefore, crucial that higher education institutions that offer education in non-mainstream languages consider the potential impact on their ability to detect unethical use of AI in these contexts.

Human Versus AI Detection

A key contribution to highlighting the need for effective AI detection testing is measuring academic staff's ability to distinguish AI-generated content. To this end, Perkins et al. (2023b) conducted a study to explore the implications of AI-generated content in academic settings, focussing on Turnitin's detection capabilities, the ability of academic staff members to evaluate detection results, and their perceptions about AI-generated content. Their sample consisted of 22 experimental submissions using GPT-4, as accessed through ChatGPT, to simulate student responses on a wide range of topics and genres within a business school environment. They found that Turnitin, although imperfect, could better identify AI generated content as opposed to human evaluators.

Comparative Analysis of Results of the AI Detection Efficiency Studies

Binary Classification Approaches

Table 1.4 provides a comparative analysis of various AI detection tools based on their performance and predictive metrics, as researched by Elkhatat et al. (2023), Bellini et al. (2024), and Popkav and Barret (2024). Each tool is evaluated for its sensitivity (true positive rate), specificity (true negative rate), positive predictive value (PPV), and negative predictive value (NPV) across different versions of AI models, specifically GPT-3.5, GPT-4, and Claude. The metrics indicate the effectiveness of each tool in accurately identifying AI-generated content versus human-written content.

Comparative analysis of AI detection tools reveals varying results. Sensitivity (TPR), the measurement for the identification of AI-generated content, varied widely among tools. Notably, Originality and OpenAI Classifier demonstrated the most consistent performance, while Crossplag and Writer were the worst performers, and GPTZero, Writer, and ZeroGPT varied the most. Specificity (TNR), indicating human-generated content was correctly identified, was consistently high for Crossplag, Originality, and Copyleaks, suggesting reliable detection capabilities. Interestingly, the OpenAI Classifier did not correctly identify human-generated content (Elkhatat et al., 2023). The Positive Predictive Value (PPV), the proportion of true positive results among all positive detections, and Negative Predictive Value (NPV), the proportion of true negative results among all negative detections, also highlighted the superior performance of Crossplag and Originality across most metrics. The variability observed across metrics and the variability in the results of measuring various GenAI tools (GPT3.5, GPT4 and Claude) is most likely due to the difference in its model configuration and training. The variation could also be due to variations in the sample methods. However, what still needs to be taken into

Table 1.4 Comparative performance metrics of AI detection tools

Criteria	Metrics	GPTZero	Writer	ZeroGPT	Crossplag	Originality	OpenAI classifier	Copyleaks
Performance	Sensitivity / true positive rate (TPR)	GPT 3.5–93% / GPT 4–27% (Elkhatat et al., 2023) GPT 4–75% (Bellini et al., 2024)	GPT 3.5–80% / GPT 4–7% (Elkhatat et al., 2023) GPT 4–0% (Bellini et al., 2024)	GPT 4–75% (Bellini et al., 2024) GPT4–16% / Claude-4% (Popkav & Barret, 2024)	GPT 3.5–1% / GPT 4–40% (Elkhatat et al., 2023) GPT 4–80% (Bellini et al., 2024)	GPT 4–100% (Bellini et al., 2024) GPT4–100% / Claude-27% (Popkav & Barret, 2024)	GPT 3.5–100% / GPT 4–87% (Elkhatat et al., 2023)	GPT 3.5–1% / GPT 4–93% (Elkhatat et al., 2023)
	Specificity / true negative rate (TNR)	GPT 3.5–80% / GPT 4–80% (Elkhatat et al., 2023) GPT 4–100% (Bellini et al., 2024)	GPT 3.5–40% / GPT 4–40% (Elkhatat et al., 2023) 100% (Bellini et al., 2024)	GPT4–50% (Bellini et al., 2024) 38% (Popkav & Barret, 2024)	GPT 3.5–100% / GPT 4–100% (Elkhatat et al., 2023)	GPT 4–100% (Bellini et al., 2024) 97% (Popkav & Barret, 2024)	0% (Elkhatat et al., 2023)	GPT 3.5–80% / GPT 4–80% (Elkhatat et al., 2023)
Predictive value	Positive predictive value (PPV)	GPT 3.5–93% / GPT 4–80% (Elkhatat et al., 2023)	GPT 3.5–86% / GPT 4–33% (Elkhatat et al., 2023)	n/a	GPT 3.5–100% / GPT 4–100% (Elkhatat et al., 2023)	n/a	GPT 3.5–75% / GPT 4–72% (Elkhatat et al., 2023)	GPT 3.5–13% / GPT 4–93% (Elkhatat et al., 2023)
	Negative predictive value (NPV)	GPT 3.5–80% / GPT 4–27% (Elkhatat et al., 2023)	GPT 3.5–67% / GPT 4–15% (Elkhatat et al., 2023)	n/a	GPT 3.5–100% / GPT 4–36% (Elkhatat et al., 2023)	n/a	GPT 3.5–0% / GPT 4–100% (Elkhatat et al., 2023)	GPT 3.5–100% / GPT 4–80% (Elkhatat et al., 2023)

consideration is that these samples are not representative of the complex textual composition found in a real academic context. Ultimately, based on the results of these studies, there is no clear indication of whether these detection methods would be feasible for the academic context.

AI Detection Under Conditions of Obfuscation and Adversarial Techniques

Table 1.5 presents performance metrics for AI text detection tools as evaluated by Weber-Wulff et al. (2023) and Perkins et al. (2024). Amongst the tools evaluated, which will be presented here include ZeroGPT, GPTZero, and Turnitin, each tool being measured across different accuracy metrics: Binary accuracy (straightforward classification as correct or incorrect), semi-binary accuracy (partial correctness), and logarithmic accuracy (gradations in correctness based on the magnitude of error). These metrics provide a comprehensive view of each tool's effectiveness in detecting AI-generated content, highlighting its strengths and weaknesses in various contexts.

The significant variance in measured performance between these two studies could be attributed to the differences in sampling methods. Weber-Wulff et al. (2023) conducted their tests using a controlled and diverse sets of simulated academic scenarios. This process included text manipulation, such as manual editing and machine translations. In contrast, Perkins et al. (2024) applied adversarial techniques to their samples entirely automated via GenAIs. The better performance, relative to Weber-Wulff et al. (2023), is likely linked to the automated adversarial techniques, which suggests that GenAI's still retain some form of predictable patterns for AI detection to detect. As stated by Perkins et al. (2024), their 'experimental design did not mirror the iterative process that is likely to occur during the use of GenAI-supported academic writing and editing, particularly if there is a deliberate attempt to evade detection.'

Table 1.5 Performance metrics of AI text detection tools (Perkins et al., 2024; Weber-Wulff et al., 2023)

Tool	Metrics	Weber-Wulff et al. (2023)	Perkins et al. (2024)
ZeroGPT	Binary accuracy	59%	40%
	Semi-binary accuracy	67%	46%
	Logarithmic accuracy	72%	52.3%
GPTZero	Binary accuracy	54%	16%
	Semi-binary accuracy	66%	24%
	Logarithmic accuracy	81%	39%
Turnitin	Binary accuracy	76%	56%
	Semi-binary accuracy	79%	62%
	Logarithmic accuracy	68%	65%

Table 1.6 Performance metrics of AI text detection tools (Perkins et al., 2024)

Tool	Mean accuracy (non-manipulated)	Mean accuracy (manipulated)	Drop in accuracy (%)
ZeroGPT	31.3%	17.3%	14%
GPTZero	26.3%	16.7%	9.7%
Turnitin	50%	7.9%	42.1%

Turnitin consistently shows the highest performance, indicating its robustness and effectiveness in detecting AI-generated content—even though it was ranked second to Copyleaks in the Perkins et al. (2024) study. ZeroGPT exhibits a varied performance with strengths in logarithmic accuracy but inconsistencies in other metrics. However, GPTZero generally underperforms compared to the other tools, showing potential in probabilistic assessments but significant challenges in binary accuracy. This analysis underscores the need for continuous improvement of AI detection tools to effectively maintain academic integrity amidst advancing AI technologies. However, as presented in Table 1.6, a comparative analysis of three AI text detection tools in terms of their performance after the application of adversarial techniques also emerges. The data highlight each tool's mean accuracy before and after the application of adversarial techniques and the percentage drop in accuracy. The metrics were derived from testing a set of AI-generated and human-written samples, emphasising the challenges and effectiveness of these tools in educational settings more like real-world scenarios.

What stands out here is the stark drop in accuracy, particularly with Turnitin. While this may seem concerning, it's important to recognise that once text generated by a GenAI tool is modified through human intervention, it is no longer 'fully AI.' Several AI detection tools are highly effective at identifying unaltered AI-generated content. However, as these studies show, their performance declines once human intervention occurs—indicating they still function as designed. This calls for a rethinking of our expectations for these tools and a clearer understanding of what exactly we need them to detect.

ANT Analysis of AI Detection as a Primary Actor

Returning to ANT analysis, I would like to demonstrate the problematics of AI detection within the higher education network. During the *problematisation stage*, arguably where higher education is at this juncture, AI detection may be seen as a probable solution. This dependence comes from the reactive tendencies created by plagiarism detection cultures. Now, with the rise of the evolving sophistication of generative AI technologies, detection systems (or strategies) are failing (Bellini et al., 2024). As these technologies produce increasingly human-like text–and tools are now specifically developed for cheating–problematisation becomes more complex. Furthermore, as demonstrated in the above literature review, AI detection in its current form cannot process the complexity and diversity of academic writing, which these tools must accurately parse and analyse, i.e., the recursive nature of academic writing (Eckstein et al., 2011). Moreover, the application of adversarial techniques, such as machine paraphrasing and obfuscation, these techniques can be both deliberate and unintentional, stemming from students' natural writing processes or their attempts to evade detection (Perkins et al., 2024). The *interessement stage* involves strategies to lock various actors into their roles, including AI detection tools. For AI detection tools, this means convincing educators, administrators, and students of their effectiveness and necessity, as well as developing tools capable of detecting AI in complex academic settings. Above, all these actors need to be trained on the nuances of AI detection once reliable methods/strategies have been developed. Educators and administrators may not be convinced of the effectiveness of tools, leading to resistance or inconsistent adoption. Additionally, and most concerning, students' awareness of detection limitations can foster attempts to evade detection, further complicating the *interessement stage*. *Enrolment* occurs when actors accept their roles and align their actions with the goals of the network. Problems in this stage include the lack of standardised evaluation metrics, which results in varied and sometimes contradictory findings about the tool's effectiveness (Elkhatat et al., 2023). This lack of consensus hinders widespread acceptance and alignment. Furthermore, discrepancies in training data lead to performance variability, limiting the effectiveness / reliability of AI detection tools. As a consequence, stakeholders cannot unify their approach. Mobilisation, the final stage, is where the solution gains widespread acceptance and becomes a black box, taken for granted by the actors involved. The primary issue here is the incomplete translation process due to the problems mentioned above.

Recommendations for Future Research: Authorship Attribution

A probable reason for the failure of AI detection as implemented now is that it views the submitted text as its own *universum,* i.e., the text as a self-contained universe. AI detection methods focus solely on the internal characteristics of the text, and these tools miss critical contextual cues that could help distinguish AI-generated content from human-written content. Hereby, I propose the development of AI detection strategies related to authorship attribution.

Authorship attribution is the task of identifying the author of a given text by analysing the unique features of their writing style (Misini et al., 2022). This involves comparing the text in question with a corpus of texts whose authorship is already known, using various linguistic and stylistic features such as word frequency, sentence structure, and syntactic patterns to determine the likelihood of authorship (Lagutina et al., 2019). Although GenAI will most likely be continuously integrated into higher education, there will still be instances where authentic writing will need to be measured. On the one hand, Bekker (2024) proposed a five-tier system to simplify thinking about permissions and prohibitions related to the use of GenAIs for academic writing. The first three tiers of Bekker are based on human-generated content: *Tier 1: Use ban* does not allow any use of AI; *Tier 2: The proofing tool* reserves GenAI after the writing process; and *Tier 3:* The copyediting tool accommodates the use of GenAI during the writing process to "shorten wordy text (e.g., reducing an abstract from 300 to 150 words), expand for clarity, and rephrase for precision are tasks every academic writer has worked on and would likely welcome assistance with. On the other hand, Furze et al. (2024) and Perkins et al. (2023a) have proposed the AI Assessment Scale (AIAS) to promote transparency and ethical use of GenAI tools. The first three levels of AIAS, such as Bekker (2024), are based on the generation of human content. In both cases, that is, the five-tier system and AIAS, the final levels or levels allow for extensive GenAI use. By implication, the first three levels or levels would require AI detection methods, as many institutions cannot afford the luxury of conducting assessments offline. For these purposes, authorship attribution via corpora containing authorship profiles could likely present a feasible solution. Although these techniques are not new, they could significantly enhance AI detection tools by capturing original student writing, allowing a comparison with subsequent submissions to detect inconsistencies or signs of AI-generated content.

Recent studies such as Huang et al. (2025) provide compelling findings on centring detection efforts around authorship attribution. As part of their systematic literature review, they identified four core problems with AI-detection: (1) human-written text attribution, (2) LLM-generated text detection, (3) attribution of texts to specific LLMs or humans, and (4) attribution of human–LLM co-authored texts. These four effectively capture different approaches to research with AI and provides a suitable framework to investigate the abilities of detection systems, and subsequent ethical frameworks. For each, the most effective methods involve transformer-based neural networks (e.g., BERT, RoBERTa) or large language models used either for feature extraction or as end-to-end attribution systems. These models outperform traditional AI detectors by capturing authorial style, adapting across domains, and providing greater transparency in their predictions. Notably, authorship attribution, according to these researchers, is the only approach capable of handling both detection (is this AI-generated?) and attribution (who or what generated this?), giving it a methodological advantage in addressing complex academic integrity cases.

To further demonstrate this position, the research of André et al. (2023) attempted to contribute to the current understanding of how specific linguistic characteristics differentiate between AI and human written texts. Their findings indicate that AI-generated text consistently produces: (1) lower perplexity levels; (2) higher grammatical correctness; (3) higher frequency of same- and longer-sequence n-gramme distributions (repeated use of longer word sequences); (4) lower type token rations (less lexical variety); (5) bias towards longer token (preference for longer words); and (6) broader frequency of function words. It follows that AI writes differently from humans.

However, in the case where language models such as Claude become more human-like or models specifically designed to mimic humans as a means of cheating, AI detection strategies could incorporate authorship attribution to capture unique writing features. In this regard, Guo et al. (2023) followed a three-phase approach in English and Chinese, namely human evaluations, linguistic analysis, and content detection experiments, to study the characteristics of ChatGPT-generated content compared to human experts. The data set for their experiment, called the Human ChatGPT Comparison Corpus (HC3) experiment, was compiled from publicly available question-answering data sets and Wiki text. Their linguistic analysis consisted of a random collection of one human answer and one from ChatGPT for each question. The study found that ChatGPT's average response length was on average longer than humans (198.14 words compared to 142.50 words for human responses). The size of the ChatGPT vocabulary was larger (79,157 compared to human size of 66,622). However, the vocabulary density of ChatGPT's responses was lower (1.41 compared to human density of 2.33). This was indicative of a less varied word usage despite the longer responses and larger vocabulary. In their parts-of-speech comparison (POS), ChatGPT's responses contained a higher frequency of nouns, verbs, adjectives, and adverbs, suggesting a tendency towards more structured and informative content, while human responses showed greater use of pronouns and conjunctions, indicating more conversational and diverse language use. Their dependency parsing analysis reveals similar characteristics in that where ChatGPT uses more determination, conjunction, and auxiliary relations. It is also no surprise that their sentiment analysis found that ChatGPT expresses more neutral sentiments. These findings suggest that while ChatGPT generates longer and more structured responses with a larger vocabulary, its language use is less varied and more neutral compared to the more conversational and diverse style of human responses. Ultimately, authorship attribution techniques, albeit not fully developed for this context, could provide a feasible solution for AI detection. This strategy could enhance our understanding of student writing and continual research in the nuances of AI detection.

Conclusion

In conclusion, the effectiveness of AI-detection tools in higher education is currently limited by various factors, including their inability to account for the broader context of academic writing and the sophisticated evasion techniques employed by users. In this chapter, ANT proved to be a valuable framework for understanding the complex interplay between the different actors involved in AI detection. By analysing these dynamic relationships between students, educators, AI technologies, and institutional policies, we can identify the strengths and weaknesses of current detection methods. The review of existing research highlights the need for continuous development and refinement of AI detection strategies to ensure they remain effective in an evolving academic landscape. Future research should focus on integrating authorship attribution methods to enhance the detection of AI-generated content, providing a more reliable means of maintaining academic integrity. By adopting a holistic approach that considers unique writing styles and contextual factors, institutions can better address the challenges posed by generative AI and foster an environment of ethical scholarship and authentic learning.

Acknowledgements I would like to acknowledge the iterative feedback provided by ChatGPT, which supported the refinement of ideas, concepts, and language. However, I take full responsibility for the interpretation, accuracy, and final presentation of all content in this work.

References

André, C. M., Eriksen, H. F., Jakobsen, E. J., Mingolla, L. C., & Thomsen, N. B. (2023, November 6–7). Detecting AI authorship: Analyzing descriptive features for AI detection. Paper submitted at the seventh workshop on natural language for artificial intelligence, Rome, Italy. https://ceur-ws.org/Vol-3551/paper3.pdf.

Bekker, M. (2024). Large language models and academic writing: Five tiers of engagement. *South African Journal of Science, 120*(1–2), 1–5. https://doi.org/10.17159/sajs.2024/17147

Bearman, M., Tai, J., Dawson, P., Boud, D., & Ajjawi, R. (2024). Developing evaluative judgement for a time of generative artificial intelligence. *Assessment & Evaluation in Higher Education, 49*(6), 893–905. https://doi.org/10.1080/02602938.2024.2335321

Bellini, V., Semeraro, F., Montomoli, J., Cascella, M., & Bignami, E. (2024). Between human and AI: Assessing the reliability of AI text detection tools. *Current Medical Research and Opinion, 40*(3), 353–358. https://doi.org/10.1080/03007995.2024.2310086

Bolanos, F., Salatino, A., Osborne, F., & Motta, E. (2024). Artificial intelligence for literature reviews: Opportunities and challenges. *arXiv*. https://doi.org/10.48550/arXiv.2402.08565

Chaka, C. (2023). Detecting AI content in responses generated by ChatGPT, YouChat, and Chatsonic: The case of five AI content detection tools. *Journal of Applied Learning and Teaching, 6*(2). https://doi.org/10.37074/jalt.2023.6.2.12

Eckstein, G., Chariton, J., & McCollum, R. M. (2011). Multi-draft composing: An iterative model for academic argument writing. *Journal of English for Academic Purposes, 10*(3), 162–172. https://doi.org/10.1016/j.jeap.2011.05.004

Elkhatat, A. M., Elsaid, K., & Almeer, S. (2023). Evaluating the efficacy of AI content detection tools in differentiating between human and AI-generated text. *International Journal for Educational Integrity, 19*(1), 17. https://doi.org/10.1007/s40979-023-00140-5

Essien, A., Bukoye, O. T., O'Dea, X., & Kremantzis, M. (2024). The influence of AI text generators on critical thinking skills in UK business schools. *Studies in Higher Education, 49*(5), 865–882. https://doi.org/10.1080/03075079.2024.2316881

Fenwick, T. J. (2010). (un)Doing standards in education with actor-network theory. *Journal of Education Policy, 25*(2), 117–133. https://doi.org/10.1080/02680930903314277

Furze, L., Perkins, M., Roe, J., & MacVaugh, J. (2024). The AI assessment scale (AIAS) in action: A pilot implementation of GenAI supported assessment. *arXiv*. https://doi.org/10.48550/arXiv.2403.14692

Guo, B., Zhang, X., Wang, Z., Jiang, M., Nie, J., Ding, Y., Yue, J., & Wu, Y. (2023). How close is ChatGPT to human experts? Comparison corpus, evaluation, and detection. *arXiv*. https://doi.org/10.48550/arXiv.2301.07597

Hass, A. (2022). Using plagiarism feedback as assessment for learning to socialise students into disciplinary writing: A theoretical perspective. *South African Journal of Higher Education, 36*(2), 133–150. https://hdl.handle.net/10520/ejc-high_v36_n2_a9.

Hosseini, M., Resnik, D. B., & Holmes, K. (2023). The ethics of disclosing the use of artificial intelligence tools in writing scholarly manuscripts. *Research Ethics, 19*(4), 449–465. https://doi.org/10.1177/17470161231180449

Huang, B., Chen, C., & Shu, K. (2025). Authorship attribution in the era of LLMs: Problems, methodologies, and challenges. arXiv:2408.08946v2 [cs.CY]. https://llm-authorship.github.io/.

Keith, K., & Van Belle, J. P. (2014, December). The use of a social networking site in the facilitation of internationalization in higher education: A case study using the actor network theory perspective. In *2014 IEEE fourth international conference on big data and cloud computing* (pp. 516–523). IEEE. https://doi.org/10.1109/BDCloud.2014.72

Lagutina, K., Lagutina, N., Boychuk, E., Vorontsova, I., Shliakhtina, E., Belyaeva, O., Paramonov, I., & Demidov, P. G. (2019, November). A survey on stylometric text features. In *2019 25th conference of open innovations association (FRUCT)* (pp. 184–195). IEEE. https://doi.org/10.23919/FRUCT48121.2019.8981504

Latour, B. (2005). *Reassembling the social: An introduction to actor-network-theory*. Oxford University Press.

Misini, A., Kadriu, A., & Canhasi, E. (2022). A survey on authorship analysis tasks and techniques. *SEEU Review, 17*(2), 153–167.

Mphahlele, A., & McKenna, S. (2019). The use of Turnitin in the higher education sector: Decoding the myth. *Assessment & Evaluation in Higher Education, 44*(7), 1079–1089. https://doi.org/10.1080/02602938.2019.1573971

Paledi, V. N., & Alexander, P. M. (2017). Actor-network theory to depict context-sensitive m-learning readiness in higher education. *The Electronic Journal of Information Systems in Developing Countries, 83*(1), 1–26. https://doi.org/10.1002/j.1681-4835.2017.tb00619.x

Páscoa, C., & Tribolet, J. (2014). Actor network theory applied to organizational change: A case study. In G. Jamil, A. Malheiro, & F. Ribeiro (Eds.), *Rethinking the conceptual base for new practical applications in information value and quality* (pp. 233–248). IGI Global. https://doi.org/10.4018/978-1-4666-4562-2.ch010

Perkins, M., Furze, L., Roe, J., & MacVaugh, J. (2023a). Navigating the generative AI era: Introducing the AI assessment scale for ethical GenAI assessment. *arXiv*. https://doi.org/10.48550/arXiv.2312.07086

Perkins, M., Roe, J., Postma, D., McGaughran, J., & Hickerson, D. (2023b). Detection of GPT-4 generated text in higher education: Combining academic judgement and software to identify generative AI tool misuse. *Journal of Academic Ethics, 22*(1), 89–113. https://doi.org/10.1007/s10805-023-09492-6

Perkins, M., Roe, J., Vu, B. H., Postma, D., Hickerson, D., McGaughran, J., & Khuat, H. Q. (2024). Simple techniques to bypass GenAI text detectors: Implications for inclusive education. *International Journal of Educational Technology in Higher Education, 21*(1), 53. https://doi.org/10.1186/s41239-024-00487-w

Pesovski, I., Santos, R., Henriques, R., & Trajkovik, V. (2024). Generative AI for customizable learning experiences. *Sustainability, 16*(7), 3034. https://doi.org/10.3390/su16073034

Popkov, A. A., & Barrett, T. S. (2024). AI vs academia: Experimental study on AI text detectors' accuracy in behavioral health academic writing. *Accountability in Research*. https://doi.org/10.1080/08989621.2024.2331757

Tatnall, A. (2009). Actor-network theory applied to information systems research. In M. Khosrow-Pour (Ed.), *Encyclopedia of information science and technology* (2nd ed., pp. 20–24). IGI Global. https://doi.org/10.4018/978-1-60566-026-4.ch004

Tatnall, A., & Burgess, S. (2002, June). *Using actor-network theory to research the implementation of a BB portal for regional SMEs in Melbourne, Australia*. Paper delivered at the 15th bled electronic commerce conference: eReality: Constructing the eEconomy, bled, Slovenia, University of Maribor, June 2002. https://domino.fov.um.si/proceedings.nsf/0/3d1df2a6b3884d1ec1256e9f00389d8a/$FILE/tatnall.pdf.

Tauginienė, L., Gaižauskaitė, I., Glendinning, I., Kravjar, J., Ojsteršek, M., Ribeiro, L., Odiņeca, T., Marino, F., Cosentino, M., Sivasubramaniam, S., & Foltýnek, T. (2018). *Glossary for academic integrity* (ENAI report 3G, revised version). European Network for Academic Integrity (ENAI). http://www.academicintegrity.eu/wp/.

Weber-Wulff, D., Anohina-Naumeca, A., Bjelobaba, S., Foltýnek, T., Guerrero-Dib, J., Popoola, O., Šigut, P., & Waddington, L. (2023). Testing of detection tools for AI-generated text. *International Journal for Educational Integrity, 19*(1), 26. https://doi.org/10.1007/s40979-023-00146-z

Chapter 2
Academic Integrity and Staff Perceptions: Perspectives on Students' Understanding of Academic Integrity, and Staff Needs in the Age of Artificial Intelligence

Caroline Campbell

Abstract This chapter reports the key findings from a staff survey undertaken at a Russell Group university in the UK in the academic years 2021–22 and 2022–23. The survey was developed as part of a Leeds Institute for Teaching Excellence (LITE) Fellowship – "Academic Integrity Strategies: positive, preventative, and punitive" – which sought to develop a positive, preventative approach to this complex area. The survey focused on obtaining staff perceptions of students' understanding of academic integrity in a taught student context; understanding current practice as regards providing guidance to students on academic integrity; ascertaining staff concerns about identifying academic malpractice in assessment; and identifying staff needs in order to inform training needs and resources. The survey also invited respondents' suggestions on how to encourage an institutional culture of academic integrity. As the availability of artificial intelligence (AI) had recently come to the fore, the 2022–23 version of the survey included some questions on this topic to elicit staff perspectives on the use of AI tools in their teaching and their views on students' use. The responses provided key insights which have informed three main recommendations: a staff training programme, resources to better support staff, and an institutional approach. The rapid development of GenAI underlines the institutional need to provide regular staff updates regarding emerging trends, to design relevant staff training, and to provide an agile response in updating policy. This should develop a shared understanding and foster greater sharing of practice across the Higher Education sector as we seek to prepare our students for their future.

Keywords Academic integrity · Higher education · Staff training needs · Institutional approach · Artificial intelligence

---

C. Campbell (✉)
University of Leeds, Leeds, UK
e-mail: c.campbell@leeds.ac.uk

I. Glendinning et al. (eds.), *Ethics and Integrity in Education (Research)*, Ethics and Integrity in Educational Contexts 9,
https://doi.org/10.1007/978-3-031-98157-9_2

Introduction

Amidst growing concerns about the threats to academic integrity and the emerging problem of contract cheating, it seemed important to deepen our understanding of the complex issues surrounding academic misconduct and explore the challenges faced by Higher Education Institutions (HEIs). As part of a Leeds Institute for Teaching Excellence (LITE) Fellowship on Academic Integrity Strategies, awarded in 2020, the project sought to develop a deeper understanding of these issues to enable the development of academic integrity strategies which would support a positive learning environment and encourage students to study with integrity (Campbell & Waddington, 2024a). In terms of the project aims, one key aim was to develop innovative, effective practice in educating students about academic integrity, and another was to identify staff needs in order to design training and better support staff awareness and understanding of academic integrity.

In each academic year between 2020 and 2023, my fellow researcher, Dr. Lorna Waddington (School of History), and I conducted an annual student survey to better understand taught students' awareness of academic integrity, to ascertain whether enough guidance was provided and to inform improvements that could be made (Campbell & Waddington, 2024b). The findings would be particularly useful to us in our School roles as Academic Integrity Lead and Assessment Lead and would provide feedback on how we, at an institutional level, could better support students in developing their understanding of academic integrity. Our attention next focussed on staff as part of our overall project aim of providing positive scaffolding for staff and students. Inspired by the Harper et al. (2019) survey of university staff in Australia, which focussed on the topic of contract cheating, we decided to conduct a staff survey to explore staff perceptions of students' awareness and understanding of academic integrity in a taught student context; to elicit current practice in terms of the support which teaching staff offered to students; to gauge staff awareness of academic integrity; and to identify any staff training needs. The responses were invaluable in providing a snapshot of practice, which we would not otherwise have had, and identified further guidance, training and resources staff thought were required. The staff survey was first conducted in 2021–22 and repeated in 2022–23.

This chapter explores the findings from the study and their implications for institutional policy and individual practice. The findings have informed our recommendations at School, Faculty, and University levels. Identifying the gap in the support for staff around academic integrity matters led to the development of staff training and a staff-facing website with resources on trends and developments. A new role for an Institutional Academic Integrity Lead was created as part of the refreshed approach to academic integrity. A new Institutional Assessment Lead role was also created, indicating the sense of priority given to these two interconnected areas. With both leads working closely together, the University benefits from their combined expertise and their familiarity with recent research and developments, and ensures that policy is regularly updated, and training is fit for purpose.

Method

We designed an anonymous online survey using Microsoft Forms. The focus was on gaining the perceptions of teaching staff and insights into their teaching practice in a taught student education context. With permission, two of the questions were similar to those of the survey undertaken by Harper et al. (2019). The 2021–22 survey contained 39 items, which were a combination of predominantly nominal scale questions (Yes / No / Don't know / N/A questions, which are quick to answer), multiple choice questions with dropdown menus, and five open questions. The first section asked about the respondent's role, their Faculty and School, how many years of teaching experience they had in Higher Education, which levels of study they taught, their gender, and whether they had experience of being an Academic Personal Tutor (APT). The second section asked about their perception of students' understanding of academic integrity and used largely Yes/No questions to ascertain the individual's approach to providing guidance to their students. The third section focussed on staff awareness of academic integrity and asked largely Yes/No questions, but included open questions to give respondents the opportunity to expand on their answers. In response to significant changes to the digital landscape and the implications for teaching, learning and assessment, the 2022–23 survey contained the same 39 items but included an additional item about "artificial intelligence tools, such as ChatGPT" – this item used a table format to ask five questions, using a three-way Likert scale of Yes/No/Don't know options. The final question was an open question which invited additional comments. A thematic analysis of the responses to the qualitative questions was undertaken to identify key themes.

Participants

As permission was not given to send an email to all staff, a convenience sampling method was used – this entailed sending an email invitation with a link to the survey via various staff distribution lists in an attempt to reach as many colleagues with responsibility for teaching as possible across the institution. The email was also circulated to contacts via the Leeds Institute for Teaching Excellence (LITE) distribution list, the LITE Team, and to the Directors of Student Education with the support of the project sponsor, the Dean for Student Education (Quality and Standards).

Survey Demographics

The 2021–22 survey ran for 10 weeks, from 21 January until 31 March 2022. The timing coincided with a period of strike action, and this may have affected the response rate. 249 responses were obtained – 94% of respondents described their

Table 2.1 Response rate by Faculty

Response rate by Faculty	2021–22 survey	2022–23 survey
Arts, humanities and cultures (AHC)	21%	27%
Biological sciences	4%	8%
Business	20%	6%
Engineering and physical sciences (EPS)	10%	19%
Environment	17%	7%
Medicine and health	15%	13%
Social sciences	9%	17%
Other	4%	3%

Table 2.2 Breakdown of Faculties by School

Faculty	Schools
Arts, humanities and cultures (AHC)	Design; English; fine art, history of art and cultural studies; history; languages, cultures and societies; media and communication; music; performance and cultural industries; philosophy, religion and history of science
Biological sciences	Biology; biomedical sciences; molecular and cellular biology
Business	Accounting and finance; economics; international business; management; marketing
Engineering and physical sciences (EPS)	Chemical and process engineering; chemistry; civil engineering; computing; electronic and electrical engineering; mathematics; mechanical engineering; physics and astronomy
Environment	Earth and environment; food science and nutrition; geography; Institute for Transport Studies
Medicine and health	Dentistry; healthcare; medicine; psychology
Social sciences	Education; law; politics and international studies; sociology and social policy
Other	Lifelong learning centre; careers; cross-disciplinary

role as 'academic', 5% as 'other' and 1% as 'teaching assistant'. The respondents came from each of the seven faculties (see Table 2.1 for a breakdown of the response rate by Faculty and Table 2.2 for a breakdown of the Faculties by School). In terms of gender, 52% identified as female, 41% as male, 6% preferred not to say, and 1% identified as non-binary. Regarding teaching experience, 28% had over 20 years of teaching experience, 33% had 11–20 years, 24% had 6–10 years, and 15% had up to 5 years of experience (see Table 2.3).

The 2022–23 survey ran for just over 4 weeks, from 17 May to 16 June 2023. This also coincided with a period of strike action, which may have affected the response rate. 142 responses were obtained, with 95% describing their role as 'academic', 2% as 'other' and 3% as 'teaching assistant'. The respondents came from all seven faculties. In terms of gender, 62% identified as female, 35% as male and 3% preferred not to say. Regarding teaching experience, 30% had over

Table 2.3 Respondents' years of teaching experience

Number of years of teaching experience	2021–22 survey	2022–23 survey
20+ years	28%	30%
11–20 years	33%	33%
6–10 years	24%	22%
Up to 5 years	15%	15%

20 years of teaching experience, 33% had 11–20 years, 22% had 6–10 years, and 15% had up to 5 years of experience.

Limitations

In terms of the limitations of this piece of research, it is acknowledged that the sample size was small (6.2% response rate based on the number of academics at the University for the 2021–22 survey and 3.5% for the 2022–23 survey). A number of factors contributed to this – for example, the researchers' request to circulate an email to all academic staff not being granted; the surveys being conducted in a background of periods of industrial action; the surveys running at a different time of year and for a different length of time for reasons beyond the researchers' control. The response rate by Faculty varies for both surveys which means any valid comparison, while potentially desirable, is not possible. Lastly, there is the potential for bias where respondents may have given socially desirable answers rather than reflecting their actual practice.

Findings

Staff Views of Students' Understanding of Academic Integrity in a Taught Student Context

We asked staff if they thought students had a good understanding of what is meant by 'academic integrity' (see Table 2.4), and responses varied. Between 50% (2021–22) and 52% (2022–23) thought that it depended on the student; between 28% and 22% said 'yes, generally'; between 14% and 12% said 'yes, by final year'; between 7% and 13% answered 'no'; and about 1% answered 'I don't know'. It is difficult to account for the increase in the proportion of 'no' responses to the 2022–23 survey except to note that this seems significant given the smaller overall response rate that year.

We asked if they thought there was enough information available to students (see Table 2.5), and about 65% of respondents to both surveys said 'Yes', on average, 18% said 'No', and, on average, 16% said 'Don't know'. Responses from those who

Table 2.4 Staff perception of students' understanding of academic integrity

Do you think your students have a good understanding of what is meant by 'academic integrity'? Please select all that apply	2021–22	2022–23
Yes, generally	28%	22%
No	7%	13%
Yes, by final year	14%	12%
It depends on the student	50%	52%
I don't know	1%	1%

Table 2.5 Staff perception of whether enough information is available to students

Do you think there is enough information available to students about academic integrity?	2021–22	2022–23
Yes, generally	66%	65%
No	17%	19%
Yes, by final year	17%	15%

answered 'No' suggested the need for formative practice, particularly at an early stage in a student's studies, and the need for regular reminders throughout the year, specifically with all summative assessments. Some respondents commented that the issue is not so much about whether there is enough information but whether students understand the need for academic integrity and are, therefore, motivated to develop their skills in this area. The importance of providing discipline-specific resources was also identified.

Asked if they thought students are given enough training on academic integrity, the response rate was very similar to both surveys, with about 37% of respondents saying 'Yes', about 37% saying 'No,' and around 26% saying 'Don't know'. Those who answered 'No' identified the need for repeated training throughout Semester 1 and beyond. The challenges faced by some international students were highlighted, and it was felt that there is a need for training and practice to be emphasised for those new to the culture of UK Higher Education. An analysis of the responses revealed the following key themes around student training:

- Adopting a supportive, positive approach;
- Engagement in understanding the principles and applying them in their work, as part of a more cohesive approach which includes how to read critically and helps students develop their own voice;
- Timely, regular training at an appropriate level with formative tasks; and,
- A broader focus to include the importance, and benefits, of demonstrating academic integrity.

Next, each of these four themes will be considered in turn.

Adopting a Supportive, Positive Approach

A recurrent theme was the need to help students understand the requirements and be able to demonstrate this in their writing and for this to be presented with a positive emphasis – "Better understanding of citation as a tool in building an academic context for argument, and not simply as a way to avoid plagiarism." There was an acknowledgement that students need to be taught what good academic practice means and be given practice in developing their academic writing – "work on how to avoid it [malpractice] i.e., looking at language and discourse to build citation rather than patch writing". One commented, "Student problems with academic integrity are not isolated but relate to their understanding more broadly of finding their individual academic voices and taking control of their discursive essay-writing."

A number of responses highlighted the tension between good intentions on the part of staff but awareness of the implications in terms of workload – "We would like to do developmental formative practice with our first years and postgraduate taught students (PGTs) over first semester, but are not workloaded for formative work, and we have classes in the hundreds." Perhaps the use of multiple-choice quizzes, for example, would be an effective way of providing feedback to students on developmental formative tasks.

Engagement in Understanding and Applying the Principles as Part of a Cohesive Approach

Several comments alluded to students seeming to understand the requirements but struggling to put them into practice. Some were in favour of using Turnitin for a draft submission or a formative task so that all taught students could learn from the report. Interestingly, over half of the respondents to both surveys indicated that giving students access to the Turnitin report for a formative piece of work is not their current practice. Several commented on the need for regular training in student-friendly language with a focus on a supportive approach – "[training] that highlights the benefits rather than the problems that can occur if it is not followed". There was awareness that the emphasis should be on the positive as students "are very anxious about getting it wrong, so the training needs to be supportive rather than scary".

One respondent highlighted the need for "more examples of appropriate summarising and paraphrasing" together with the principles of academic integrity, as this is one of the reasons "why international students are so over-represented in academic integrity case numbers". Similarly, making time in the curriculum to discuss academic integrity and its application in the discipline was recognised as hugely beneficial – "There needs to be opportunities for students to ask questions and raise concerns, as well as learn about the behaviours which promote good academic integrity".

Timely, Regular Training at an Appropriate Level

Several responses highlighted the need for training to be introduced at appropriate times over the course of the academic year, and to be delivered on a regular basis. There was a sense that a 'one size fits all' approach does not meet student needs. Instead, a more tailored approach which builds on student understanding should be designed, appropriate to each level of study. The challenge here is postgraduate study where students have a much shorter time in which to become acquainted with academic integrity requirements.

One suggestion included Schools having "at least one seminar per year per level on academic integrity". This would give Schools flexibility in terms of the content (tailored to the discipline), the method(s) of delivery, the timing of the delivery, and appropriateness to the level of study. I would argue that one seminar per semester should be the minimum requirement.

The following comment about the current training – "content is too mechanistic and detached from their reality" - is a reminder of the need for Faculties/Schools to provide disciplinary examples and indicates the limitations of providing generic examples. One response went much further by arguing that students would benefit from undertaking training prior to commencing their studies – "A compulsory pre-commencement module on academic integrity and the purpose and practices of research". This would probably need to be a compulsory requirement for it to be taken seriously, with multiple attempts allowed and a deadline for completion, all of which would need to be tracked. The institution would need to decide the outcome if a student failed to complete the training. If such a measure were introduced, it would certainly send a message that the institution is serious about supporting students in developing their understanding of academic integrity.

The need for tailored support for students from widening participation backgrounds was highlighted – "the knowledge and practices we have are not self-evident, and we can encourage students to learn, but we also need to reflect on how we can redress inequitable access to prior knowledge." This is a reminder of the need for staff to clarify the expectations for all our students, similar to our responsibility to unpack the hidden curriculum (QAA, 2022) in an attempt to dismantle the known barriers to success.

Several thought that workshops and one-to-one sessions would be useful. There was recognition that students need to be guided through examples and not simply signposted to resources, but there was no indication of who should deliver these workshops and how these could be incorporated into already busy timetables. I would argue that University libraries are well placed to provide support and deliver such training (Campbell et al., 2024) - ideally there should be an introduction before students commence their studies, and this would continue with further training delivered at specific times throughout the student journey.

Broadening the Focus of Training

There was also feedback on the need for the current online training to have a broader focus and go beyond 'avoiding plagiarism' and to include research integrity. Indeed, one particularly pertinent comment noted the need for academic integrity to be situated "as the underlying principle of thinking and research, not as an additional requirement".

Several responses mentioned collusion and, in particular, students not understanding what is meant by collusion — "[students] need much more about collusion. Perhaps training needs to be widened out". This was also mentioned with regard to online assessments. Campbell et al. (2024) make the case that collusion should be included as part of any academic integrity training, and the authors highlight the role of university libraries in providing training.

A Snapshot of Current Practice

We asked whether they provide guidance or training about academic integrity to students in their teaching, and 78% of respondents (2021–22) and 75% (2022–23) said that they do (see Table 2.6). When this is cross-referenced with their years of teaching experience, it suggests that the more experienced educators realise the benefit of including guidance as part of an educative approach to their teaching. It is possible that this approach reduces the incidence of unintentional plagiarism, which is a positive outcome for the student and for the person marking their work.

We asked whether they provide an opportunity for students to ask questions about academic integrity in their lectures/seminars/tutorials, and 79% of respondents (2021–22) and 83% (2022–23) said that they do.

Asked whether they spend class time teaching students how to reference, 62% of respondents (2021–22) and 58% (2022–23) said 'yes' (see Table 2.7). When this is cross-referenced with the years of teaching experience, it indicates that about one third of those respondents with more than 11 years of teaching experience include the mechanics of referencing in their teaching. While there is work to be done to increase this figure, it is clear that those with less teaching experience need to be persuaded of the value of devoting class time to teaching students

Table 2.6 Percentage of respondents who include guidance, and number of years' teaching experience

Do you include guidance on academic integrity in your teaching? Yes	2021–22	2022–23
20+ years of teaching experience	28%	34%
11–20 years	35%	33%
6–10 years	21%	24%
Up to 5 years	16%	9%

Table 2.7 Percentage of respondents who spend class time teaching referencing skills and number of years' teaching experience

Do you spend class time teaching students how to reference? Yes	2021–22	2022–23
20+ years of teaching experience	29%	34%
11–20 years	33%	35%
6–10 years	25%	17%
Up to 5 years	13%	13%

Table 2.8 Categories of staff who might be considered responsible for ensuring an understanding of academic integrity

Whose responsibility is it to ensure students have a good understanding of academic integrity? Please select all that apply	2021–22	2022–23
Teaching staff	72%	73%
Academic Personal Tutors	45%	45%
Academic Integrity Leads	65%	64%
The School	76%	82%
The University	76%	75%
Other	15%	15%

how to reference and the benefits that this should bring, namely students developing their understanding of the requirements and being able to apply this to their work.

We asked whether they discuss the University's academic integrity policy with their students and the consequences of breaching it, and 70% said that they do.

The 2021–22 survey accepted responses under 'other' in relation to the question about whether they ensure students understand what is meant by academic integrity, and this revealed some interesting insights. Some respondents pointed out that it is not possible to "ensure" students understand–point taken. A small number commented that they talk about it with students if they are asked, or they highlight it in feedback if the student's work indicates poor practice. Others mentioned that they signpost students to useful resources, but this usually happens in their Academic Personal Tutorials and not in their teaching. A very small number of respondents indicated that they felt this was not their responsibility and/or they assumed it was covered in other modules.

Whose Responsibility Is It?

We wanted to pinpoint who staff members think should have responsibility for ensuring that students have a good understanding of what is required in terms of academic integrity. Respondents could choose from a pre-populated list of five options, and they could indicate 'other' (see Table 2.8). The most popular choices were 'the School', then 'the University', followed by 'teaching staff', 'Academic Integrity Leads', and 'Academic Personal Tutors' (APTs). This would suggest that the responsibility has to be shared between the institution providing policy and guidance, those in a School role providing further staff guidance and support, those in teaching roles with responsibility at a modular level, and APTs at an individual level with their tutees. The perception that it is a shared responsibility is positive.

Staff Training and Support on Academic Integrity

We asked whether there is enough guidance and/or training on academic integrity for staff, and responses were divided with 44% (2021–22) and 43% (2022–23) replying 'Yes', 36% (2021–22) and 41% (2022–23) replying 'No', and 20% (2021–22) and 16% (2022–23) replying 'Don't know'. Asked what they would like to see, the responses fell broadly into three categories – staff training, staff support and an institutional approach.

Staff Training

Responses about what training staff would like to have, included:

- requiring compulsory online training to ensure a minimum level of understanding about what academic integrity is and the University's expectations;
- workshops on academic integrity with a focus on the benefits of academic integrity for students, how to help them avoid malpractice and how to support them develop their skills, using examples of good practice from across the disciplines;
- academic integrity training as part of induction for new staff;
- training on the effective use of Turnitin; and
- guidance on how to incorporate academic integrity into teaching, and how to integrate AI tools into teaching, learning and assessment.

Staff Support

- Responses highlighted a need for the following support:
- encouraging academic staff to ask for guidance and support with identifying malpractice;
- greater clarification to ensure a better understanding of conventions and practices;
- resources, including examples of materials and good practice; and
- guidance for APTs when their tutees find that they are under investigation for academic integrity breaches.

An Institutional Approach

The following responses related to institutional-level support:

- a greater focus on support for students and an emphasis on preventative approaches;
- clarification of the expectations of the role of the APT and of teaching staff in offering students support on academic integrity;
- more discussion and guidance around the application of policy to benchmark standards and ensure greater consistency;
- more realistic work-loading of marking to reflect the time required when malpractice is suspected; and
- clarification around the use of AI tools in teaching, learning and assessment, including discussion of the opportunities, challenges, risks and concerns.

With increasing student numbers and the time and effort required to identify academic misconduct, it would make sense for this to be reflected in the workload for staff involved in marking. At the very least, this would acknowledge the time required when malpractice is suspected.

Staff Concerns Regarding Identifying Malpractice

A number of respondents took the opportunity to express particular concerns. A key concern was the amount of time required when a suspicion of malpractice is identified. On making known their concern, the onus is on the academic to find the original sources. If the student is called to a meeting to discuss the concern, this requires that the academic is also available. As one response indicated, "The time required to follow-up is considerable". One or two respondents hinted that this in itself was a deterrent to following the expected procedure – "The way in which this process generates additional work for academic and SES staff is a strong disincentive to take potential plagiarism cases further." As previously stated, including a specific allocation in the workload model for identifying and following up malpractice is critical. Another response indicated unease with the emphasis being punitive rather than taking an educative approach – "[the] main concern is that by identifying it, the system is then set up to punish rather than educate, support and guide". In a similar vein, another comment explained why an academic might feel disinclined to take the matter further – "The process is long, arduous and stressful for both the staff and student. It often feels like an accusation has been made and so the relationship between students and staff can be damaged as a result". This highlights the need for a review of the process, particularly in the light of shifting to an educative approach. The option of using Turnitin as a learning tool for a formative piece of work could be beneficial here but would need to be applied at an institutional level so that all students have this opportunity.

Table 2.9 Responses to questions on Artificial Intelligence tools at a glance

Artificial Intelligence tools	Yes	No	Don't know
Have you used AI tools, such as ChatGPT?	53%	47%	0%
Have you considered how AI tools could be used in your discipline?	73%	25%	2%
Do you see AI as a largely positive development?	28%	36%	36%
Do you support student use of AI tools?	31%	38%	31%
Do you think staff training on the use of AI tools would be useful?	92%	4%	4%

Responses Regarding Artificial Intelligence (AI) Tools

The 2022–23 survey asked five questions on the topic of AI tools using a three-way Likert scale of Yes/No/Don't know (see Table 2.9). The responses provide a useful insight into staff perceptions.

- 53% of respondents had used AI tools such as ChatGPT.
- 73% had considered how they could use AI tools in their discipline.
- Responses to the question of whether they saw AI as a largely positive development were divided, with 28% saying 'Yes', 36% saying 'No', and 36% saying 'Don't know'.
- Asked whether they supported student use of AI tools, responses were more evenly divided, with 31% saying 'Yes', 38% saying 'No', and 31% saying 'Don't know'. Those saying 'Yes' were from each Faculty, with the highest responses of 45% and 44%, respectively, from staff in Biological Sciences and Engineering and Physical Sciences (EPS), followed by 31% from Arts, Humanities and Cultures (AHC), 25% from Social Sciences, 22% from Business, 20% from Environment, and 17% from Medicine and Health.
- Asked whether they thought staff training on the use of AI tools would be useful, 92% said 'Yes', representing each Faculty, while 4% said 'No' and 4% said 'Don't know'. Those responding 'No' were from AHC, Biological Sciences, Medicine and Health, and Social Sciences; those responding 'Don't know' were from AHC, Biological Sciences, EPS and Social Sciences.

Discussion

The findings from the 2021–22 survey and the 2022–23 survey provide staff perceptions of students' understanding about academic integrity, a snapshot of current practice and staff concerns, and identify specific training and resources that would support them in developing their awareness and enhancing their practice. The 2022–23 responses offer insights into the complexity of teaching, learning and assessment in the light of the rapid development of artificial intelligence tools. Bretag (2013) argued that everyone has responsibility for ensuring that students develop a good understanding of what academic integrity is and how it should be demonstrated, and this is reflected in the QAA's principles for academic integrity (2020). Marsh and Campion (2018) identified the gap between what staff expected students to know and their actual prior knowledge and argued for a collaborative approach involving teaching staff, learning advisors and library staff. In our role as educators, we need to move away from any assumption that students will or should know what is required and instead take responsibility for developing their understanding and their ability to demonstrate academic integrity by including it in our teaching. We need to provide opportunities in our teaching for students to learn what the disciplinary requirements and conventions are in a UK HE environment. It could be argued that it makes sense to reduce the amount of subject content in order to cover the relevant academic integrity requirements since students need to understand this as part of their university education. Kier and Ives' (2022) study identified the need for resources and training to support staff in delivering this effectively. Curtis et al.'s (2021) evaluation of a series of workshops confirmed their effectiveness in increasing staff awareness and confidence. These need to be themed and run on a continuous basis in a hybrid manner to maximise staff engagement.

Eaton (2021) argues that upholding academic integrity requires a multistakeholder approach which involves everyone from students, educators, those who support student learning, administrators through to senior management, the institution itself and beyond, into the community. Chugh et al. (2021) and Luck et al. (2021) remind us that 'all' teaching staff includes part-time staff, those on short contracts, and those employed as teaching assistants, many of whom are often postgraduate research students or students completing their PhD. It is vital that they also have a good understanding of academic integrity requirements and include this in their teaching. At the University of Leeds, where the research was conducted, there is currently no specific time allocation devoted to "training", a point that was highlighted by one respondent – "Training is not included in the workload model. It should be included so that TAs [Teaching Assistants] and hourly paid staff can be paid for training and for permanent staff to have a specific time allocation for it."

Ideally, all staff who teach compulsory modules should include discussion in their teaching, appropriate to the level of study, about what academic integrity is and how it should be demonstrated in the discipline. Similarly, academic integrity training should be embedded in all modules on postgraduate programmes as postgraduate students need a fast-track approach. Campbell and Waddington (2024b) maintain that it is the module leader's responsibility to ensure that the discipline-specific requirements pertaining to academic integrity are included in the module. Providing teaching materials for staff, such as slides, which can be edited, and links to resources that staff can refer to, would assist with this task. It is good practice to include a formative academic integrity task in the first semester to give students the opportunity to put their understanding into practice and to gain feedback (Bretag et al., 2018; Morris, 2016). The feedback might be in the form of seeing their Turnitin report for this formative assessment. Building in opportunities for self-evaluation and for peer feedback would develop student autonomy and would shift the focus from a reliance on tutor feedback. Students can be further supported by attending pre-sessional and/or in-sessional classes, where understanding academic integrity requirements, and the opportunity to practise meeting those requirements, is a key part of the syllabus.

Staff also wanted to have more opportunities for "informed discussions about the complexities of academic integrity, malpractice and plagiarism, drawing on empirical studies of influencing factors". This suggests the need for those with responsibility for academic integrity, such as the Institutional lead and/or the School lead, to be the expert and to provide regular updates on trends and developments. This would align with the reflections of Curtis et al. (2021) on the need for workshops to be delivered by staff who have expertise in the topic. As already mentioned, discipline-specific guidance is important too.

A key institutional challenge is delivering appropriate training to staff and obtaining staff buy-in. Including an allocation for training and professional development in the workload model would show the institution's commitment to this. Some annual online training at the University of Leeds, such as Health and Safety, is compulsory, so it would seem appropriate to add academic integrity for those who teach. Perhaps having an 'academic integrity-themed' week in the academic calendar, to coincide with the International Day of Action for Academic Integrity, would encourage discussion between staff and students and would further promote a culture of academic integrity. The TEQSA toolkit (Lodge et al., 2023) includes links to resources designed by various institutions, including Monash University's module for staff on supporting academic integrity in teaching. Workshops could make staff aware of other useful resources, such as the London and South East Academic Integrity Network's Contract cheating detection checklist (LSEAIN-CCWG, 2021), and could provide updates on emerging threats.

Artificial Intelligence (AI) Tools

Clearly the availability of AI and the emergence of generative AI tools has developed rapidly since the 2022–23 survey was conducted, but it is interesting to gauge how staff felt about the subject at the time of the survey (May/June 2023). The invitation to comment further on the use of AI tools attracted 73 responses. A common theme was that, as educators, we should be proactive in supporting students in learning how to use the tools effectively – "I don't think we can stop the use of these tools, and so it is vital that we teach students how to use the tool in terms of the use of sources, cohesion, critical thinking, structure and addressing the question." Another commented, "We should be integrating activities now into our syllabi, which incorporate explicit activities around what AI can and cannot be useful for – not only for academic writing, but also study life in general." This approach is in line with the Russell Group (2023) principles on the use of generative AI tools in education, which underline the shared commitment to ensuring AI literacy for both students and staff. As Lodge et al. (2023) point out, "it will become increasingly important to consider how these tools are integrated into learning and teaching in higher education in intelligent ways" (2023, p. 2). Building staff capability has workload implications, and this needs to be addressed.

In terms of assessment, some respondents were already fully aware of the implications and commented that tasks should be designed with AI tools in mind – "I will try and include this in assessment in a way that encourages positive use of AI tools". Similarly, "It is pervasive, so better to design means to assess with that in mind than to attempt to regulate it as it is unlikely to be effective use of our time and capability". Another felt that it would be acceptable for students to use AI "as long as there were elements ... which required students to verify their knowledge (e.g. by viva)". The QAA (2023) guidance stresses the need for institutions to support students in developing their artificial intelligence literacy and to have clear policies in place. Lodge et al. (2023) provide guiding principles and propositions which can guide discussions and decisions as institutions review assessment practices in the light of AI. Waring and Evans (2024) confirm the need for institutional guidance and training so that educators can encourage students' critical evaluation of AI tools. This forms an important part of their "principled approach" to the integration of GenAI in assessment and feedback. Dawson (2023) explores using GenAI to scaffold learning and deciding when it is useful for students to use it e.g. depending on whether it is intrinsic to the assessment learning outcomes. One respondent commented on the need for accreditation bodies to be aware of the potential impact of AI tools – "Accreditation bodies need to be made more aware of this as currently our assessment practices are limited as to what we can do". Several commented on the challenge of detecting whether AI tools had been used and their concerns about how to handle this. Other respondents focussed on the long-term impact on assessment and felt that a return to in-person exams was the way forward, together with a move to oral assessment in the form of progression interviews. Clearly the implications of AI tools for assessment are a significant cause for concern for some staff, and further research is needed to address this. The Irish National Academic Integrity Network (NAIN, 2023) guidelines stress the importance of providing training for staff and students and that "appropriate resources are allocated to this area to ensure that all avail of it" (2023, p. 9).

A number of respondents seemed to be waiting for further guidance and institutional policy to clarify the extent to which students would be allowed to use AI in general and GenAI in particular for assessment purposes and were hoping that training would be forthcoming. The need for clarification in the form of clear policy and guidance is supported by recent studies which have obtained students' perceptions of GenAI in higher education (Chan & Hu, 2023; Johnston et al., 2024). In the spirit of partnership, it is important to consult students and create the guidance together.

Lodge et al. (2023) assert that, "There will be a widespread need to build academic and professional staff capabilities for working with AI" (2023, p. 7). It is a poignant reminder that academic integrity should not be treated as a standalone entity – it is closely connected with assessment, institutional values and, ultimately, the reputation of the award. Dawson (2023) comments that "GenAI is currently the least capable it will ever be" (2023, time 20:16), and we need to identify the opportunities for integrating it into our teaching and learning as well as address the risks. As Eaton (2023) argues, we need to "prepare students for a future we cannot yet imagine – their future, not ours" (2023, p. 9).

Conclusion

The findings from the institutional staff survey run in two academic years (2021–22 and 2022–23) offer insights into staff perceptions and provide feedback on the training and support that would assist them in their role as educators, particularly given the challenge of integrating AI tools into teaching, learning and assessment. It is clear from the findings and from recent literature that institutional investment is urgently needed in order to provide training and resources to meet staff needs as they develop their awareness and enhance their practice. At an institutional level, there is a need for all staff with responsibility for teaching to engage with some element of compulsory training on academic integrity with the aim of ensuring a baseline of knowledge. Regular updates and refresher activities will be required if institutions are to demonstrate their commitment to staff – and students—in terms of delivering not just policy but also good practice around assessment design, identifying academic misconduct, on-going awareness of developments, emerging threats and opportunities, and working with students to educate, support and guide in a digital age. As ever, we, as educators, need to be informed ourselves so that we can effectively support our students. As we navigate the everchanging environment, working smartly, responding quickly, and continuing to share strategies and practice across the HE sector, all relevant parties need to work together to address the constant challenge of anticipating trends and developments and understanding their implications. Future research might focus on discussions between students and staff about better defining the role of AI in teaching, learning and assessment.

Acknowledgements The author would like to thank: the Leeds Institute for Teaching Excellence (LITE) for supporting the LITE Fellowship for 3 years; her co-researcher, Dr. Lorna Waddington; the late Dr. Tracey Bretag, and Prof. Rowena Harper for giving permission to use their survey questions; those colleagues who made time to respond to the staff survey and whose invaluable responses have informed this research and led to important developments in policy, resources and training; Anne Stazicker for providing feedback on the draft.

References

Bretag, T. (2013). Challenges of addressing plagiarism in education. *PLoS Medicine, 10*(12), e1001574. https://doi.org/10.1371/journal.pmed.1001574

Bretag, T., Harper, R., Burton, M., Ellis, C., Newton, P., & Rozenberg, P. (2018). Contract cheating: A survey of Australian university students. *Studies in Higher Education, 44*(11), 1837–1856. https://doi.org/10.1080/03075079.2018.1462788

Campbell, C., & Waddington, L. (2024a). Leeds institute for teaching excellence fellowship: Academic integrity strategies. University of Leeds. https://teachingexcellence.leeds.ac.uk/research/fellowships/academic-integrity-strategies-positive-preventative-protective-and-punitive/.

Campbell, C., & Waddington, L. (2024b). Academic integrity strategies: Student insights. *Journal of Academic Ethics, 22*, 33–50. https://doi.org/10.1007/s10805-024-09510-1

Campbell, C., Haikney, E., & Waddington, L. (2024). Collusion: The new norm? In J. Seeland & J. Openo (Eds.), *Academic integrity and the role of the academic library: Institutional examples and promising practices* (pp. 139–155). Springer. https://doi.org/10.1007/978-3-031-65731-3_8

Chan, C. K. Y., & Hu, W. (2023). Students' voices on generative AI: Perceptions, benefits, and challenges in higher education. *International Journal of Educational Technology in Higher Education, 20*(43). https://doi.org/10.1186/s41239-023-00411-8

Chugh, R., Luck, J. A., Turnbull, D., & Pember, E. R. (2021). Back to the classroom: Educating sessional teaching staff about academic integrity. *Journal of Academic Ethics, 19*(1), 115–134. https://doi.org/10.1007/s10805-021-09390-9

Curtis, G., Slade, C., Bretag, T., & McNeil, M. (2021). Developing and evaluating nationwide expert-delivered academic integrity workshops for the higher education sector in Australia. *Higher Education Research & Development, 41*(3), 665–680. https://doi.org/10.1080/07294360.2021.1872057

Dawson, P. (2023). Don't fear the robot: Future-authentic assessment and generative artificial intelligence [Keynote address, video]. YouTube. https://www.youtube.com/watch?v=ZEhRFVO6rr0.

Eaton, S. E. (2021). *Plagiarism in higher education: Tackling tough topics in academic integrity*. Bloomsbury Publishing. https://doi.org/10.5040/9798400697142

Eaton, S. E. (2023). Postplagiarism: Transdisciplinary ethics and integrity in the age of artificial intelligence and neurotechnology. *International Journal for Educational Integrity, 19*(23). https://doi.org/10.1007/s40979-023-00144-1

Harper, R., Bretag, T., Ellis, C., Newton, P., Rozenberg, P., Saddiqui, S., & van Haeringen, K. (2019). Contract cheating: A survey of Australian university staff. *Studies in Higher Education, 44*(11), 1857–1873. https://doi.org/10.1080/03075079.2018.1462789

Johnston, H., Wells, R. F., Shanks, E. M., Boey, T., & Parsons, B. N. (2024). Student perspectives on the use of generative artificial intelligence technologies in higher education. *International Journal for Educational Integrity, 20*(2). https://doi.org/10.1007/s40979-024-00149-4

Kier, C. A., & Ives, C. (2022). Recommendations for a balanced approach to supporting academic integrity: Perspectives from a survey of students, faculty and tutors. *International Journal for Educational Integrity, 18*(22). https://doi.org/10.1007/s40979-022-00116-x

Lodge, J. M., Howard, S., Bearman, M., Dawson, P., & Associates. (2023). Assessment reform for the age of artificial intelligence. Tertiary Education Quality and Standards Agency. https://www.teqsa.gov.au/sites/default/files/2023-09/assessment-reform-age-artificial-intelligence-discussion-paper.pdf.

LSEAIN-CCWG. (2021). *Contract cheating detection for markers: Checklist of strategies and features commonly found in assignments produced through contract cheating*. https://sites.google.com/view/contractcheatingwglseain/our-checklist.

Luck, J. A., Chugh, R., Turnbull, D., & Pember, E. R. (2021). Glitches and hitches: Sessional academic staff viewpoints on academic integrity and academic misconduct. *Higher Education Research and Development, 41*(4), 1152–1167. https://doi.org/10.1080/07294360.2021.1890697

Marsh, J. D., & Campion, J. (2018). Academic integrity and referencing: Whose responsibility is it? *Journal of Academic Language and Learning, 12*(1), A213–A226. https://journal.aall.org.au/index.php/jall/article/view/546.

Morris, E. J. (2016). Academic integrity: A teaching and learning approach. In T. Bretag (Ed.), *Handbook of academic integrity* (1st ed., pp. 1037–1053). Springer. https://doi.org/10.1007/978-981-287-098-8_11

National Academic Integrity Network. (2023). Generative artificial intelligence: Guidelines for educators. QQI. https://www.qqi.ie/sites/default/files/2023-09/NAIN%20Generative%20AI%20Guidelines%20for%20Educators%202023.pdf.

QAA. (2020). *Academic integrity charter for UK higher education*. QAA. https://www.qaa.ac.uk/docs/qaa/about-us/academic-integrity-charter.pdf?sfvrsn=93f0d181_8

QAA. (2022). Unpacking your hidden curriculum: Guide for educators. QAA. https://www.qaa.ac.uk/docs/qaa/members/unpacking-your-hidden-curriculum-guide-for-educators.pdf?sfvrsn=51d7a581_8.

QAA. (2023). Maintaining quality and standards in the ChatGPT era: QAA advice on the opportunities and challenges posed by generative artificial intelligence. QAA. https://www.qaa.ac.uk/docs/qaa/members/maintaining-quality-and-standards-in-the-chatgpt-era.pdf.

Russell Group. (2023). Russell group principles on the use of generative AI tools in education. https://www.russellgroup.ac.uk/sites/default/files/2025-01/Russell%20Group%20principles%20on%20generative%20AI%20in%20education.pdf.

Waring, M., & Evans, C. (2024). Facilitating students' development of assessment and feedback skills through critical engagement with generative artificial intelligence. In C. Evans & M. Waring (Eds.), *Research handbook on innovations in assessment and feedback in higher education: Implications for teaching and learning* (pp. 330–354). Elgar Publishing. https://doi.org/10.4337/9781800881600.00027

Chapter 3
An Approach to Evaluate the Impact of Academic Integrity Training for Educators by Analysing Assessment Referrals

Lauren Schrock, Asima Iqbal, and Alexa Kirkaldy

Abstract To maintain academic integrity in education, there is a need to translate policies and regulations into practices of the learning community. In the institution that is the subject of this chapter, this occurs in a department where a Director of Academic Integrity leads strategic initiatives for ensuring consistency in academic integrity. One initiative is the evaluation of academic integrity training for educators by analysing their referrals for possible misconduct in student work. This is important to providing an evidence-based approach to academic integrity training to ensure regularity and quality in pedagogical practice. Therefore, the chapter investigates the question, "How to evaluate the impact of academic integrity training for educators on their referrals of student assessments for potential academic misconduct?". In answer to this question, the chapter shares the development of a database containing educators' referrals of student work for possible cases of academic misconduct to propose how this data will be used to evaluate the impact of academic integrity training for educators. While a work-in-progress, the chapter provides practical guidance on the collection of referral data and proposes ways in which this data may be analysed to inform training. Hence, the work contributes to the development of a method for collecting and analysing assessment referrals that can support academic integrity leadership to make strategic decisions on support for educators in the practice of academic integrity.

Keywords Academic integrity · Academic integrity strategy · Academic integrity training · Academic misconduct referrals · Analysing academic misconduct referrals · Referrals database

---

L. Schrock (✉) · A. Iqbal · A. Kirkaldy
WMG, University of Warwick, Coventry, UK
e-mail: Lauren.Schrock@warwick.ac.uk; Asima.Iqbal@warwick.ac.uk; A.M.Kirkaldy@warwick.ac.uk

I. Glendinning et al. (eds.), *Ethics and Integrity in Education (Research)*, Ethics and Integrity in Educational Contexts 9,
https://doi.org/10.1007/978-3-031-98157-9_3

Introduction

While there is no fixed meaning to 'academic integrity' (Macfarlane et al., 2014), there are many definitions that may inform institutional policy (e.g. International Center for Academic Integrity, 2021; Tauginienė et al., 2018). As a result, understanding of academic integrity and its counterpart academic misconduct are contextually derived. For instance, universities create their own policies and regulations to identify and classify misconduct and subsequent processes for their investigation and consequences (Möller, 2023). This can explain confusion and differences in the practice of academic integrity since what is acceptable in one institution may not be in another.

To support educators to carry out their responsibility for academic integrity in alignment with the institution, it is important to provide training. Training can ensure educators are consistent in their approach to integrity, for example, identifying different forms of misconduct and referring possible cases of misconduct for independent investigation in accordance with institutional policy and regulation (Eaton, 2021). Also, training is beneficial to support educators to respond to emerging issues, such as the development and proliferation of generative artificial intelligence tools (Cotton et al., 2024). While academic integrity training for educators can be assessed by self-reported change, there is an opportunity to examine how data on the decision-making of misconduct can be collected and analysed to ensure misconduct is approached consistently (East, 2009). This is strategically important for leadership in academic integrity since such evidence can inform educator training to promote a culture of academic integrity.

This chapter addresses the question, "How to evaluate the impact of academic integrity training for educators on their referrals of student assessments for potential academic misconduct?" by examining a department within a Russell Group University (The Russell Group of Universities, n.d.). In this department, a Director of Academic Integrity leads the development of a database of assessment referrals for student work as well as the design and delivery of academic integrity training for educators. These initiatives, therefore, are the subject of the chapter. At the time of writing, the collection of referrals for an entire academic year is incomplete, so the chapter is a work-in-progress. Nonetheless, the chapter contributes to wider practice by proposing a method for collecting and analysing assessment referrals as an evidence-based approach to evaluate academic integrity training for educators. This is beneficial for academic integrity leadership in guiding educators to be consistent and accurate in academic integrity since this is of strategic significance for student satisfaction, educator workloads, and institutional reputation.

Literature Review

Academic integrity is the responsibility of students, educators, and researchers in the learning community of the university (International Center for Academic Integrity, 2021). Educators can compromise academic integrity by providing incorrect guidance to students (Malik et al., 2021) and failing to identify or report instances of misconduct, so misconduct remains unaddressed (Satalkar & Shaw, 2018). When academic integrity is undermined, this can negatively impact a university's credibility and reputation (Heuser et al., 2016) and student wellbeing (Eaton et al., 2023). One way to avoid these consequences is academic integrity training for educators.

There are several learning outcomes of academic integrity training for educators that are important for maintaining a culture of academic integrity. One is to help educators identify forms of misconduct in student work. Due to the contextual nature of misconduct, terminology of misconduct may be elicited from institutional policy (Resnik et al., 2015). Therefore, developing a shared understanding for different forms of misconduct and how they may appear in assessment is important so educators are consistent when recognising instances of misconduct. Such consistency is beneficial so resolutions to misconduct can be tailored to the specific occurrence. For example, advice to students on how to avoid plagiarism can differ to guidance on avoiding the fabrication of data. It is important to note that as the educational environment evolves due to advancements in artificial intelligence and access to contract writers, educators must be informed about how these changes influence the identification of misconduct (Morris, 2018). Hence, academic integrity training for educators must be revised and re-delivered as changes impact the determination of misconduct in student work.

Another learning outcome of academic integrity training is to ensure educators follow integrity processes and policies of the department and university (Kier & Ives, 2022). One important process is the reporting of possible misconduct by educators (Dannhoferová et al., 2022). This is important so student work suspected of misconduct receives a fair and consistent outcome. Hence, training can involve guidance on how to raise a "plagiarism incident report" for a potential case of academic misconduct exhibited in student work (Chugh et al., 2021, p. 127). Such reports are significant to assisting the investigation of academic misconduct in that they identify the alleged type(s) of misconduct with supporting evidence. Hence, training that provides "clarity about evidence required" can strengthen the rigour of the referral process (Kier & Ives, 2022, p. 6).

An additional learning outcome of academic integrity training is for educators to be able to use text similarity detection tools. Tools like Turnitin (2024) can be helpful for inspecting text similarity (e.g., plagiarism, collusion), yet the determination of potential academic misconduct relies on the educators' judgement (Sutherland-Smith & Carr, 2005). This is an important distinction since reliance on tools for detecting text similarity to police and punish academic misconduct can lead to an over-simplification of integrity issues and missed pedagogic opportunities (Mphahlele & McKenna, 2019). Consequently, training on text similarity detection tools benefits educators to make their own academic judgement rather than relying solely on the tool to determine misconduct.

Due to the importance of the learning outcomes of academic integrity training, training must be available to all educators. For example, educators on temporary contracts may be less aware about the regulations, procedures, and student support that informs the landscape of academic integrity since they may be less integrated into the learning community due to their conditions of work compared to educators with a full-time contract (Hudd et al., 2009). For instance, educators known as 'externals', 'associates', or 'sessional teachers' may be managing employment at other institutions which have separate, potentially conflicting, guidance on academic integrity. Therefore, it is important that educators who are temporarily employed in an institution receive training on academic integrity as practiced in the institutional setting to ensure consistency in responding to academic misconduct, as this benefits student learning (Chugh et al., 2021; Luck et al., 2022). For similar reasons, it is important that educators who are experienced in academic integrity yet new to the institution, as well as educators with primarily industry experience, also receive training. This is because training can ensure an alignment amongst educators with institutional regulations and processes and competency in using text similarity detection tools.

The evaluation of academic integrity training is important to evidence good practice. However, conducting an evaluation of such training can be a challenge (Andrews & Glendinning, 2024). One approach to assess training is through self-reported evaluations completed by participants post-training. In a study conducted by Curtis et al. (2021), participants completed surveys to assess workshops aimed to provide guidance on promoting academic integrity, responding to breaches, and fostering a culture of integrity within higher education institutions; findings of the study revealed the workshops led to a significant increase in participants' awareness of key academic integrity issues and the size of the institution had a significant impact on the changes in participants' confidence related to academic integrity practices. This, according to the authors, was because in larger institutions, communicating policies and practices was deemed to be more difficult, with a potential of hindering staff awareness of all academic integrity initiatives that were implemented (Curtis et al., 2021). Therefore, academic integrity training can play a crucial role in enhancing awareness and confidence regarding academic integrity among higher education staff. However, the perception of integrity may contrast with behaviour, hence "[a]nother way to promote consistency of decision making is to collect and map data from decision makers and lecturers about how they react to suspected cases of academic misconduct" (East, 2009, p. A46-A47). This is the opportunity the chapter aims to address.

Hence, the chapter proposes a method for collecting and analysing referrals made by educators for student work evidencing potential misconduct. The findings can be used by academic integrity leadership to evaluate academic integrity training for educators. This is strategically significant to ensure educators are consistent and accurate in the practice of academic integrity.

Academic Integrity Training for Educators and Referral of Academic Misconduct

The chapter focuses on a department within a Russell Group university in which a Director of Academic Integrity oversees the strategic implementation of pedagogical practices and academic procedures to enhance the academic integrity of educators and students in alignment with university policies and regulations. A strategic aim of the Director is to ensure educators are consistent in the practice of academic integrity. This is paramount to ensuring educational credibility and high-quality student learning and assessment.

To promote consistency in educators' practice of academic integrity in the assessment of student work, the Director leads two initiatives. First is the design and delivery of academic integrity training for educators. The Director oversees the creation of training materials, so content comprises several sources: institutional published regulations and guidance, departmental resources, collective expertise from individuals in the department with significant experience in academic integrity, references to pedagogical scholarship and links to guidance on how to use a text similarity detection tool. Content is updated each academic year to advise educators of changes. Since academic integrity is a strategic priority in the department, attendance at training is mandatory for full-time staff and educators on temporary contracts are paid a nominal fee to promote attendance and engagement. Training, to be accessible to educators, is provided as an online workshop. A recording of the workshop is also available for educators to review. This academic integrity training for educators is regularly scheduled: every 3 months, there is a session for new starters, and twice-a-year, there is a training session delivered prior to a student's dissertation submission. This ensures that educators, whether full-time or temporary staff, with assessment responsibilities, undertake this training.

By participating in the workshop, it is expected that educators will be able to do the following:

1. Appreciate the difference between 'poor academic practice' and 'academic misconduct' as defined in the university regulation on academic integrity;
2. Identify different types of poor academic practice and academic misconduct;
3. Recognise potential indicators relating to contract writing and use of artificial intelligence within student work;
4. Use the university-supported text similarity detection tool in conjunction with academic judgment to interpret whether student work maintains academic integrity or contains evidence of poor academic practice or potential academic misconduct; and
5. Appropriately respond to instances of poor academic practice or potential cases of academic misconduct within assessment marking or use of the referral process.

While there can be a lack of consensus on the extent to which educators are responsible for instructing students on academic integrity (Löfström et al., 2015), the workshop emphasises that educators have a vital role in reminding, teaching, and demonstrating the behaviours that reflect the institution's academic integrity values.

The workshop also clarifies that academic integrity has evolved since the rise of artificial intelligence, which means universities are revisiting their policies and regulations (Moorhouse et al., 2023). Therefore, this training is updated in accordance with the principles on artificial intelligence in education published by The Russell Group of Universities (2023).

A second strategic action led by the Director is improvement of the referral process of possible academic misconduct in student work. Similar to other universities (e.g., Mahmud, 2024; Möller, 2023; Stoesz & Eaton, 2022), underpinning the department's practice of academic integrity is a university regulation on academic integrity. In regard to the subject of this chapter, the university regulation (University of Warwick, 2022) has three key functions: delineates academic misconduct from poor academic practice, defines types of academic misconduct, and outlines a process for investigating possible cases of misconduct. Since the department decides how cases are referred for investigation, an online referral form is used to collect information from an educator about an alleged case of possible misconduct. Following the launch of ChatGPT and the proliferation of generative artificial intelligence tools and their disruption of academic integrity (Sullivan et al., 2023), in March 2023, the online referral form was amended to include artificial intelligence as a reason for referral. An outline of the referral form is presented in Table 3.1.

Table 3.1 Referral form for possible academic misconduct in student assessment

Section	Explanation
Essential information	The referrer must provide information about themselves (e.g., staff ID number), university email address (to receive a copy of the referral), and information about the student whose work is being referred (e.g., student ID number, program of enrolment) The referrer must select from a list the type of issue (or issues) they are reporting. The list includes types of academic misconduct, poor academic practice, and the use of artificial intelligence as the reason(s) for referral To support the investigation, the referrer is asked to identify where the problem is located (e.g., in a student's coursework, dissertation, examination or 'elsewhere')
Justification for referral	The referrer must provide a description of evidence for the referral, such as specific page numbers and/or paragraphs within the assessment and percentage of text similarity. It is noted that educators are taught in training not to rely on the percentage of similarity alone, as high similarity does not always mean there is a problem, and low similarity does not always mean there is a lack of a problem. Hence, educators apply their academic judgement to treat each assessment individually in preparation of a referral. In cases when suspected misconduct is not evidenced in a written assessment, referrers are asked to clearly describe the issue A qualitative response to justify the referral enables a referrer to provide as much detail as possible and flexibility to draw on different forms of evidence (e.g., weblinks to sources that have a high similarity with the student's work) in support of their academic judgement
Further information	The referrer can upload an additional document in support of the referral (e.g., a text similarity report), provide an ID number for another student for concerns about potential collusion, and select from a list of indicators for potential use of artificial intelligence

Relevant to this chapter is the accumulation of referral forms into a database for analysis. At the time of writing, referral data has been collected for two partial academic years since issuing the amended form in March 2023. It is difficult to compare the data collected so far from the 2022–2023 and 2023–2024 academic years due to the impact of developments in artificial intelligence on training and guidance for educators. Hence, referral data will be analysed after the conclusion of the academic year (September 2024) to obtain a holistic perspective on the impact of academic integrity training for educators for a full academic year. This analysis is significant for academic integrity leadership to evaluate educators' decision-making on academic integrity to strategically inform training based on referral behaviours.

A Method for Collecting and Analysing Assessment Referrals to Evaluate Academic Integrity Training for Educators

While a work-in-progress, the development of a method for collecting online referrals for possible misconduct in student assessment and analysing these referrals is beneficial to academic integrity leadership. This is because findings can evidence the impact of training on educators' behaviours for recognising and justifying student work for possible academic misconduct. Hence, the section proposes quantitative and qualitative avenues for analysing assessment referrals to evaluate academic integrity training for educators in support of practice and recognise the limitations of the database for future work.

In the design of the online referral form, there are several questions that lend themselves to quantitative analysis. For example, in collecting essential information of the referral it is possible to determine the frequency in which a specific form of misconduct is referred. Analysing whether certain types of misconduct are referred more frequently than others may provide evidence for academic integrity leadership where there may be gaps in practice (e.g., edge cases in which there are a few or extreme number of referrals). The frequency of different forms of misconduct can be specified further by educators and programs of study to elicit comparisons of misconduct referrals across subject teams and levels of study and identify where there may be variations in practice. This is important for academic integrity leadership to ensure educators are consistent in referrals of misconduct to benefit all students in their learning and assessment.

On collecting a significant volume of referrals into the database, it is possible to conduct more advanced statistical analysis since there is a potential for a large data set given the number of students and the multiple assessments they complete during their studies. For instance, in the essential information, an educator is asked to identify where the alleged misconduct has occurred. Hence, with a data set of significant size it may be possible to investigate whether there is a relationship in the form of student assessment (e.g., essay, presentation) and identified type of misconduct. This is beneficial for academic integrity leadership to determine if certain types of misconduct are more prevalent in specific forms of assessment. As a result, academic integrity training for educators can be tailored to support good practice: samples of student work selected for training can be intentionally picked from those referred and found to exhibit the form of misconduct commonly occurring in that type of assessment. In addition, training can be enhanced to support assessment design so educators can be proactive to avoid forms of misconduct occurring in student work.

In a justification for the referral, educators have the flexibility of an open question to provide as much evidence as needed to support the referral. Hence, this is appropriate for qualitative analysis. One approach is to code the different types of evidence used to support the referral (e.g., percentage of text similarity, examples of misconduct in student work, or external sources) and analyse the frequency in which each type of evidence appears among the collected referrals. This analysis can be further specified by analysing the frequency in which certain evidence is drawn on to support an alleged type of misconduct. For academic integrity leadership, this analysis is beneficial to inform academic integrity training since understanding common forms of evidence used to underpin academic judgement can elicit guidance for educators on good practice.

In addition to analysing the academic judgement of the referral, it is also possible to assess the alignment between the justification for the referral and the type of issue selected in essential information. Hence, the mixing of quantitative and qualitative data is beneficial to analyse the coherency of an educator's referral. This mixing is important for academic integrity leadership to shape training to ensure educators have a deep knowledge for different forms of misconduct.

However, there are limitations to this method in development. First, the database collects *referrals* rather than *outcomes* of the referral. For example, the current database does not include an outcome from the investigation for misconduct (this is a constraint of the database rather than a procedural concern). As a consequence, there is a concern about the quality of the referral data to represent *actual* misconduct (e.g., is there a case of misconduct for the student to answer to) and *accurate* identification of misconduct (i.e., a misalignment between the type of misconduct alleged in the referral compared to the type found in an investigation). To address this limitation, it is necessary to expand the capabilities of the database and include the outcomes of misconduct investigations in planning the data analysis. When the quantitative analysis of the referrals is compared to the outcomes of investigations, this can enable academic integrity leadership to determine the frequency of incorrect judgements regarding misconduct to enhance training.

Second, the database does not capture data from educators that do not complete a referral. A lack of a referral can mean that the student work has no misconduct, or the work has unidentified misconduct and is not referred, or misconduct is addressed outside of the referral process. Of strategic importance to academic integrity leadership is ensuring student work that does contain possible misconduct is referred; therefore, additional initiatives to support educators in their responsibility for academic integrity are needed. For instance, it would be beneficial to check if a student whose work is referred for misconduct at the end of an academic year also has evidence of possible misconduct in earlier assessments that were not referred. Another strategic aim of academic integrity leadership is ensuring misconduct is addressed consistently. Therefore, further data collection may be needed to investigate misconduct addressed outside of the referral process in the marking and moderation of student assessments.

As a work-in-progress, this method for collecting and analysing assessment referrals to evaluate academic integrity training for educators is significant for academic integrity leadership. This is because the quantitative and qualitative analysis of referrals can be used to evaluate the extent to which learning outcomes of training are met in the referral behaviours of educators. In addition, the analysis can inform training by recognising gaps in practice that may require further support. Hence, the analysis is important for academic integrity leadership to determine where limited time and resources may be best allocated to inform training to support the consistency and accuracy in educators' academic judgement of misconduct in student work.

Discussion

A contribution of the chapter is to propose how referral data for possible cases of misconduct can be collected and analysed to evidence the impact of and recognise improvements for, academic integrity training for educators, which can facilitate a consistent approach in the determination of misconduct (East, 2009). To consider the potential implications of this work-in-progress, this section explores the potential for ongoing monitoring of referrals and the online referral form in relation to literature. This work-in-progress is also discussed amongst larger themes of student experience and professional development of educators as pertinent to a culture of academic integrity.

A key component of the referral process is the online referral form. Since a purpose of training is to support educators to follow a process for reporting possible misconduct (Chugh et al., 2021; Kier & Ives, 2022), it is important that attention is given on how to articulate justification of possible misconduct, such as advising good practice for demonstrating sound judgement through evidence. It is also important to consider the design of the form itself since educators can prefer to address student misconduct on their own to avoid administrative burdens (MacLeod & Eaton, 2020). Academic integrity leadership must design a form that is easy-to-use, quick-to-complete, and proactive in reducing future time burdens. For example, to limit possibilities for human error, the online form used in the department discussed in this chapter has forced responses so certain items cannot be left blank, and responses are in a required format (e.g., email address). Hence, a benefit of this online referral form is to reduce unnecessary work to follow up with educators for incomplete referral forms and to ensure that the form remains up-to-date for all users. Furthermore, it is also important the online form is crafted with intent for analysis: questions must be carefully designed to ensure they elicit responses that can be analysed to support the strategic aims of academic integrity leadership. For instance, ensuring responses to the online referral form can be used to evaluate the impact of academic integrity training for educators.

As this is a work-in-progress, the approach for analysing data from the referrals is positioned as a point in the future. However, once results are found, there is an opportunity to continuously monitor the database of referrals to determine interventions in time. This is a strategic opportunity for academic integrity leadership to ensure training and guidance is designed and delivered to address burgeoning concerns and recognise good practice as they arise in the ongoing monitoring of the database. As a result, in addition to informing educators of institutional changes and developing concerns in academic integrity apparent in the educational environment (Morris, 2018; Curtis et al., 2021), training can be more timely to address academic integrity practices of educators as they occur during the academic year. This means academic integrity leadership can impact educators in their practice of academic integrity to benefit student academic integrity as students learn.

At the center of academic integrity is the student experience. It is important that referral of student work and training of educators be in the interest of supporting student learning rather than punishment. Hence, calls for compassion in academic integrity processes include recommendations for common policies that ensure a consistency to approach and training of educators to "prevent each professor from establishing their own rules and becoming, many times, judge and jury (prosecutor)" (Guerrero-Martínez et al., 2023, p. 97). This has informed the development of a database of assessment referrals to promote accountability and transparency to the identification and justification of misconduct in student work. While the current work presents specific initiatives to support educators in academic integrity, the data obtained in referrals can also be analysed to support pedagogical practice. For example, findings on the frequency in which different types of misconduct appear in different forms of assessment can be used to create student-centred resources. Providing students with guidance on avoiding common types of misconduct in specific forms of assessment can help students maintain the integrity expected of them.

Academic integrity training for educators is relevant for professional development. The principles of academic integrity translate to research integrity, which is significant for helping educators undertake rigorous pedagogical scholarship in their field. Furthermore, academic integrity training helps educators respond to changes in the educational landscape, such as the rise of generative artificial intelligence tools that impact the identification and detection of academic misconduct (Kumar et al., 2024). Hence, technical upskilling in the use of text similarity detection tools and developing a knowledge of artificial intelligence are significant aspects of integrity training that can enhance the professional competencies of educators to perform successfully in higher education. Realising the importance of sustained engagement with integrity training also supports higher education institutions to cultivate a culture of academic integrity (Curtis et al., 2021). Therefore, evaluating training through an analysis of referrals for possible misconduct is important for academic integrity leadership to support educators to grow in their profession.

In summary, the chapter shares a method in development for evaluating training by analysing assessment referrals. The quantitative and qualitative results of this analysis can be used to enhance training, guidance, and resources to promote good practice of academic integrity amongst educators and can be used to support student learning and assessment. These are significant for academic integrity leadership in building a culture of academic integrity.

Conclusion

The chapter presents a method for collecting data on referrals of student work for alleged misconduct into a database that will be analysed to evaluate the impact of academic integrity training for educators. By sharing how data will be qualitatively and quantitatively analysed, the chapter demonstrates how such results can be used to inform improvements to integrity training, such as tailoring training to improve its relevance. This training is of strategic importance for academic integrity leadership to ensure a consistent approach for misconduct that is integral to maintaining a culture of academic integrity.

There are several recommendations to support educators and institutions seeking to collect and analyse data on referrals to inform strategic actions benefitting academic integrity. One recommendation is to begin with the design and deployment of a referral form to ensure it collects data that will support an evaluation of training. To support this objective, it is important the referral form is aligned with the terminology of institutional policy and regulations. Also, educators must be informed on how to complete the form, which can be accomplished by training. When creating a database of referrals, it is recommended to assess the system's capabilities to ensure all relevant data is collected. For example, a limitation of the work-in-progress is that the database does not include outcomes of investigations for misconduct; hence, future work includes expanding the database to include additional data sources. Proactively designing the database by considering the full process of academic misconduct investigations can help ensure the database is holistic in its view to enable academic integrity leadership to address where there may be inconsistencies in practice. Finally, it is recommended that academic integrity leadership and institutions plan for how referral data will be analysed. While this chapter has proposed ways in which data will be analysed to evaluate training, it also recognises that such data can be used to develop guidance for students on maintaining academic integrity. Therefore, there are many possibilities for the analysis of referrals to assist academic integrity leadership in achieving their strategic aims. Thus, conceptualising a purpose and plan for analysing referral data can maximise its benefit.

While a work-in-progress, there are opportunities for future research once data from referrals is collected and analysed. For example, when educators are advised how data from the database has been used to enhance integrity training, it would be beneficial to examine how this may impact educators' attitudes towards leadership and the institution in its responsibility for academic integrity. Also, in advancing this work beyond supporting educators in their responsibility for academic integrity, further research can explore how the collection and analysis of referral data can be used to improve student academic integrity. This is also of strategic importance for academic integrity leadership in building a culture of academic integrity.

Overall, the chapter provides a practical contribution in sharing a method in development for collecting and analysing assessment referrals to evaluate academic integrity training for educators. This evidence can be used to inform academic integrity training and enable academic integrity leadership to achieve a strategic aim for ensuring consistency and accuracy in educators' referral of student work for possible misconduct. This is pertinent for institutions determined to enhance student experience, pedagogical practice, and credibility through the development of a culture of academic integrity.

Acknowledgements The authors would like to thank the editors for their feedback that helped to improve this chapter.

References

Andrews, S., & Glendinning, I. (2024). Governing academic integrity: Conceptualizing the assurance and efficacy of strategies and outcomes. In S. E. Eaton (Ed.), *Second handbook of academic integrity* (pp. 1081–1112). Springer.

Chugh, R., Luck, J. A., Turnbull, D., & Pember, E. R. (2021). Back to the classroom: Educating sessional teaching staff about academic integrity. *Journal of Academic Ethics, 19*, 115–134. https://doi.org/10.1007/s10805-021-09390-9

Cotton, D. R., Cotton, P. A., & Shipway, J. R. (2024). Chatting and cheating: Ensuring academic integrity in the era of ChatGPT. *Innovations in Education and Teaching International, 61*(2), 228–239. https://doi.org/10.1080/14703297.2023.2190148

Curtis, G. J., Slade, C., Bretag, T., & McNeill, M. (2021). Developing and evaluating nationwide expert-delivered academic integrity workshops for the higher education sector in Australia. *Higher Education Research & Development, 41*(3), 665–680. https://doi.org/10.1080/07294360.2021.1872057

Dannhoferová, J., Foltýnek, T., Dlabolová, D., & Fishman, T. (2022). Cutting the vicious circle: Addressing the inconsistency in teachers' approaches to academic integrity breaches. *Journal on Efficiency and Responsibility in Education and Science, 15*(1), 10–22. https://doi.org/10.7160/eriesj.2022.150102

East, J. (2009). Aligning policy and practice: An approach to integrating academic integrity. *Journal of Academic Language and Learning, 3*(1), A38–A51. https://journal.aall.org.au/index.php/jall/article/view/66.

Eaton, S. E. (2021). Academic integrity and literacy. In K. Masika (Ed.), *University teaching in focus* (pp. 163–178). Routledge.

Eaton, S. E., Pethrick, H., & Turner, K. L. (2023). Academic integrity and student mental wellbeing: A rapid review. *Canadian Perspectives on Academic Integrity, 5*(2), 34–58. https://doi.org/10.11575/cpai.v5i2.73748

Guerrero-Martínez, L. I., Ayala-Enríquez, P., & Guerrero-Dib, J. (2023). The role of compassion in academic integrity management processes. In A. Beck & A. Stengel (Eds.), *Ethics and integrity in teacher education* (pp. 89–106). Springer. https://doi.org/10.1007/978-3-031-16922-9_7

Heuser, B. L., Martindale, A. E., & Lazo, D. J. (2016). Strategic internationalization in higher education: Contexts, organizations, and implications for academic integrity. In T. Bretag (Ed.), *Handbook of academic integrity* (pp. 1–15). Springer. https://doi.org/10.1007/978-981-287-079-7_60-1

Hudd, S. S., Apgar, C., Bronson, E. F., & Lee, R. G. (2009). Creating a campus culture of integrity: Comparing the perspectives of full-and part-time faculty. *The Journal of Higher Education, 80*(2), 146–177. https://doi.org/10.1080/00221546.2009.11772137

International Center for Academic Integrity. (2021). The fundamental values of academic integrity. 3rd ed. International Center for Academic Integrity. https://academicintegrity.org/images/pdfs/20019_ICAI-Fundamental-Values_R12.pdf

Kier, C. A., & Ives, C. (2022). Recommendations for a balanced approach to supporting academic integrity: Perspectives from a survey of students, faculty, and tutors. *International Journal for Educational Integrity, 18*, 22. https://doi.org/10.1007/s40979-022-00116-x

Kumar, R., Eaton, S. E., Mindzak, M., & Morrison, R. (2024). Academic integrity and artificial intelligence. In S. E. Eaton (Ed.), *Second handbook of academic integrity* (pp. 1583–1596). Springer International Handbooks of Education. https://doi.org/10.1007/978-3-031-54144-5_153

Löfström, E., Trotman, T., Furnari, M., & Shephard, K. (2015). Who teaches academic integrity and how do they teach it? *Higher Education, 69*, 435–448. https://doi.org/10.1007/s10734-014-9784-3

Luck, J. A., Chugh, R., Turnbull, D., & Rytas Pember, E. (2022). Glitches and hitches: Sessional academic staff viewpoints on academic integrity and academic misconduct. *Higher Education Research & Development, 41*(4), 1152–1167. https://doi.org/10.1080/07294360.2021.1890697

Macfarlane, B., Zhang, J., & Pun, A. (2014). Academic integrity: A review of the literature. *Studies in Higher Education, 39*(2), 339–358. https://doi.org/10.1080/03075079.2012.709495

MacLeod, P. D., & Eaton, S. E. (2020). The paradox of faculty attitudes toward student violations of academic integrity. *Journal of Academic Ethics, 18*(4), 347–362. https://doi.org/10.1007/s10805-020-09363-4

Mahmud, S. (2024). Transformative change in academic integrity policy at Australia universities. In S. E. Eaton (Ed.), *Second handbook of academic integrity* (pp. 825–839). Spring International Handbooks of Education. https://doi.org/10.1007/978-3-031-54144-5_119

Malik, M. A., Mahroof, A., & Ashraf, M. A. (2021). Online university students' perceptions on the awareness of, reasons for, and solutions to plagiarism in higher education: The development of the AS&P model to combat plagiarism. *Applied Sciences, 11*(24), 12055. https://doi.org/10.3390/app112412055

Möller, A. (2023). An analysis of university academic integrity policies in New Zealand. *Journal of Further and Higher Education, 47*(3), 338–350. https://doi.org/10.1080/0309877X.2022.2130195

Moorhouse, B. L., Yeo, M. A., & Wan, Y. (2023). Generative AI tools and assessment: Guidelines of the world's top-ranking universities. *Computers and Education Open, 5*, 100151. https://doi.org/10.1016/j.caeo.2023.100151

Morris, E. J. (2018). Academic integrity matters: Five considerations for addressing contract cheating. *International Journal for Educational Integrity, 14*, 15. https://doi.org/10.1007/s40979-018-0038-5

Mphahlele, A., & McKenna, S. (2019). The use of Turnitin in the higher education sector: Decoding the myth. *Assessment & Evaluation in Higher Education, 44*(7), 1079–1089. https://doi.org/10.1080/02602938.2019.1573971

Resnik, D. B., Rasmussen, L. M., & Kissling, G. E. (2015). An international study of research misconduct policies. *Accountability in Research, 22*(5), 249–266. https://doi.org/10.1080/08989621.2014.958218

Satalkar, P., & Shaw, D. (2018). Is failure to raise concerns about misconduct a breach of integrity? Researchers' reflections on reporting misconduct. *Accountability in Research, 25*(6), 311–339. https://doi.org/10.1080/08989621.2018.1493577

Stoesz, B. M., & Eaton, S. E. (2022). Academic integrity policies of publicly funded universities in Western Canada. *Educational Policy, 36*(6), 1529–1548. https://doi.org/10.1177/0895904820983032

Sullivan, M., Kelly, A., & McLaughlan, P. (2023). ChatGPT in higher education: Considerations for academic integrity and student learning. *Journal of Applied Learning & Teaching, 6*(1), 31–40. https://doi.org/10.37074/jalt.2023.6.1.17

Sutherland-Smith, W., & Carr, R. (2005). Turnitin.com: Teachers' perspectives of anti-plagiarism software in raising issues of educational integrity. *Journal of University Teaching and Learning Practice, 2*(3), 106–114. https://doi.org/10.53761/1.2.3.10

Tauginienė, L., Gaižauskaitė, I., Glendinning, I., Kravjar, J., Ojsteršek, M., Ribeiro, L., Odiņeca, T., Marino, F., Cosentino, M., Sivasubramaniam, S., & Foltýnek, T. (2018, October 26). *Glossary for academic integrity*. ENAI Report 3G. https://www.academicintegrity.eu/wp/glossary/.

The Russell Group of Universities. (n.d.) *About*. https://russellgroup.ac.uk/about/.

The Russell Group of Universities. (2023, July 4). *New principles on use of AI in education*. https://russellgroup.ac.uk/news/new-principles-on-use-of-ai-in-education/.

Turnitin. (2024). *Similarity*. https://www.turnitin.co.uk/products/similarity/.

University of Warwick. (2022, September 30) *Reg. 11 Academic Integrity (effective after 4 October 2021)*. https://warwick.ac.uk/services/gov/calendar/section2/regulations/academic_integrity/.

Chapter 4
From Healthcare to Higher Ed: A Five-Level Framework for Fostering Academic Integrity

Jessica Kalra

Abstract Drawing inspiration from the five levels of preventive medicine, this chapter presents a novel framework designed to comprehensively address academic misconduct and promote a culture of academic integrity within the higher education classroom. This framework moves beyond a reactive approach to misconduct, instead advocating for a proactive and holistic strategy that tackles the root causes of misconduct while implementing effective interventions. The first tier of this preventative model, primordial prevention, emphasises critically examining systemic factors that might contribute to misconduct, such as the academy's traditional focus on the individual over the community or achievement over learning, and subsequently encourages systemic reform to create an environment that fosters academic integrity. Primary prevention focuses on cultivating a culture of integrity in the classroom, promoting ethical conduct through open communication, building relationships and raising awareness of and promoting responsible scholarship. Secondary prevention involves identifying and addressing potential risks for misconduct before it occurs and subsequently developing curricular and or pedagogical approaches to mitigate this risk. Tertiary prevention focuses on restorative practices, recognising that misconduct may occur despite preventative measures. This tier prioritises addressing violations in a manner that promotes student learning and reintegration into the academic community. When misconduct occurs, the focus is shifted away from punishment towards offering support and encouraging students to reflect on their actions and develop strategies to maintain academic integrity in the future. Finally, quaternary prevention would act to ensure that implemented measures of prevention, at any level, are unbiased and not inadvertently causing harm. Through the cyclical review and refinement of policies and practices related to the promotion of academic integrity, institutions can ensure that their efforts to promote academic integrity remain relevant, effective, and ethically sound. By integrating these five

J. Kalra (✉)
University of British Columbia, Vancouver, BC, Canada
e-mail: jessica.kalra@ubc.ca

I. Glendinning et al. (eds.), *Ethics and Integrity in Education (Research)*, Ethics and Integrity in Educational Contexts 9,
https://doi.org/10.1007/978-3-031-98157-9_4

tiers, institutions can reduce the burden of misconduct on all stakeholders, promote student engagement, and ultimately enhance learning outcomes.

Keywords Levels of prevention · Academic integrity · Academic misconduct · Social constructivism assessment design

Introduction

Preventative approaches for reducing the incidence and prevalence of academic misconduct are certainly not new. Both institutions of higher education (HE) and individual educators have been focused on defining and adopting strategies to prevent academic misconduct for decades (Bretag, 2016; Eaton, 2024; Eaton & Christensen Hughes, 2022; Rettinger & Bertram Gallant, 2022). However, much of what has been recommended with respect to prevention strategies are top down or isolated approaches that do not address the root challenges that drive academic misconduct behaviours. While some of these approaches have shown success, it is widely accepted that a multi-layered, holistic system of prevention is still needed (Macdonald & Carroll, 2006). Herein, one such system is presented.

Jason Stephens (2019) argues that "academic dishonesty" is a complex issue involving individual, psychological, situational, and cultural factors; therefore, preventing academic misconduct requires building a culture of academic integrity, a process that necessitates a comprehensive and holistic effort (Stephens, 2019). Stephens' (2016, 2019) work is compelling as it equates academic dishonesty to an epidemic and draws on the work of Cohen and Swift (1999), who write about the "Spectrum of Prevention" in the context of injury. Using the epidemic analogy, Stephens (2016, 2019) observes that allied health practitioners need a comprehensive, multilevel, and systemic framework to implement in order to see success in injury (illness and disease) prevention. He suggests that, similarly, educators need a complex framework from which to draw practical interventions that prevent academic misconduct. To support this, in 2016, Stephens introduced the "Multilevel Intervention Model," which describes a holistic approach to developing a culture of integrity.

Taking the epidemic analogy further and viewing academic misconduct through a preventative medicine lens, considerable analogous patterns between preventing academic misconduct and preventing disease, illness and injury emerge. While academic misconduct is not a pathology, it is endemic and something that educators and institutions of education want to mitigate. Reducing the risk, incidence, prevalence and impact of academic misconduct is of utmost importance to the academy. Therefore, it seems reasonable that educators use the robust, evidence-based tools, frameworks, practices and knowledge of implementation from preventative medicine as a starting point to develop relevant frameworks that aid in reducing the impact of academic misconduct and simultaneously promote academic integrity.

This chapter explores how the five levels of prevention framework from medicine can be translated for use to mitigate academic misconduct and promote academic integrity. Due to the success of the preventative framework in shifting the paradigm of health, the philosophy and language of the domains of medicine and healthcare are leveraged to facilitate this translation. The five levels of prevention; primordial, primary, secondary, tertiary, and quaternary, are first described and detailed. Subsequently, evidence-based approaches from education are mapped onto the concepts of primary, secondary, and tertiary prevention. The critical importance of primordial and quaternary prevention is highlighted, and direction is provided on how these types of approaches can be applied in HE. Finally, lessons learned from the implementation of preventative medicine, as well as the importance of a multiple stakeholder approach, are presented.

Rethinking the Definition of Academic Integrity. What Can We Learn from the Redefining of Health?

In order to define prevention strategies that reduce academic misconduct and promote academic integrity, it is important to spend time examining what integrity means contextually and subsequently position misconduct within this construct. Using the five levels of prevention framework to develop practical tools for preventing misconduct in the classroom requires a critical examination of the ontological and epistemological assumptions that are made about academic integrity as it plays out in the current and future landscape of education. Medicine has also had to confront the challenge to modernise the definition of health. Healthcare is moving away from a disease focused, paternalistic and compliance-based model of prevention to one that emphasises health promotion while at the same time recognising that the concept of health needs to be robust and relevant to an evolving landscape of medicine. According to the most recent definition from the World Health Organization (WHO), "health is a state of complete physical, mental and social well-being and not merely the absence of disease or infirmity" (Public Health, 2021). Similarly, academic integrity is not merely the absence of misconduct. However, somewhat problematically, the terms integrity and misconduct are often used as opposites; one has integrity and, therefore, does not commit misconduct and vice versa. Like the concept of health, the meaning of integrity is greater in magnitude than what happens in a classroom setting and dependent on broad social and cultural determinants. Furthermore, when considering the definition of health provided by the WHO, it is apparent that the concept of disease prevention is tightly coupled with health promotion. Part of the goal in medicine is to prevent disease/illness/injury; however, an interconnected but larger goal embedded within these preventative strategies is the goal to promote health. Put another way, prevention of disease is a relatively acute goal, while health promotion supports a long-term acquisition of the tools and mindset that enables individuals to maintain a healthful life (Green, 1979), a more chronic effect, to once again borrow the language from medicine. It is, therefore, fitting that health promotion is commonly defined as a process for enabling people to take control over and improve their health (Public Health Agency of Canada, 2001).

Figure 4.1 illustrates the bidirectional relationship between the goals of disease prevention and health promotion, indicating that preventing illness supports a

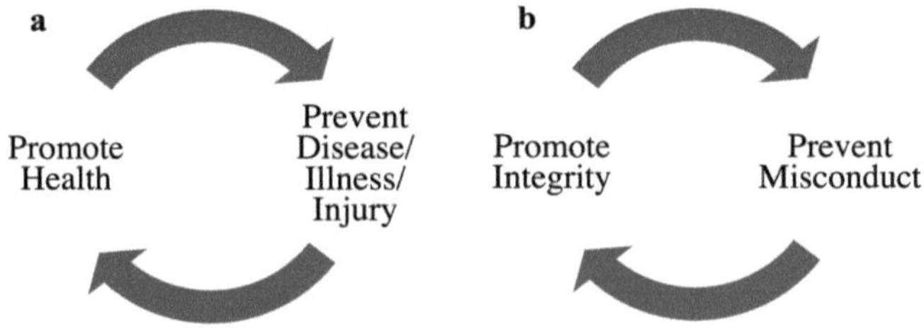

Fig. 4.1 Prevention and promotion

healthy life and healthy living decreases the risk of illness (Fig. 4.1a). Analogously, the goal in education is to prevent academic misconduct acutely and promote integrity chronically (Fig. 4.1b), both goals acting to complement each other.

The Five Levels of Prevention Framework

Preventative medicine encompasses all of the efforts and activities directed at reducing the burden of disease, illness, and injury. In addition to treatment, preventative medicine also includes reducing the risk of disease and reducing or even preventing complications (Dowdle, 1998). Preventative medicine has two arms: public health care and clinical prevention. They work in tandem to provide both population-wide and individualised approaches. The prevention framework from medicine is informed by a robust set of evidence-based practices that support these arms.

Figure 4.2 illustrates the main categories of prevention activities as they are laid out currently. The original three-tiered framework of primary, secondary, and tertiary prevention (Fig. 4.2: indicated in grey) has existed for nearly five decades (Institute of Medicine, 1980). In this form, researchers and healthcare workers employ actions within the primary, secondary, and tertiary levels of healthcare interventions, which are directed at promoting health, preventing disease, and managing existing health conditions (AbdulRaheem, 2023; Bentzen, 2003; Kisling & M Das, 2023).

Primary prevention refers to interventions that aim to prevent the onset of disease, illness, or injury. These strategies focus on promoting health and reducing the modifiable risk factors associated with various health conditions. Primary prevention interventions are generally implemented at the population level and aim to reduce the overall burden of diseases (Baumann & Ylinen, 2020; Kisling & M Das, 2023). The activities in primary prevention fall into two categories: 1. Activities and interventions that promote health and aim to enhance the overall health and well-being of the general population (Yongu, 2022). The knowledge and autonomy provided to the public using this form of prevention empowers people, giving them self-efficacy in choosing their own path towards health (Green, 1979); and 2. Activities and interventions that serve as protective measures aiming to prevent the

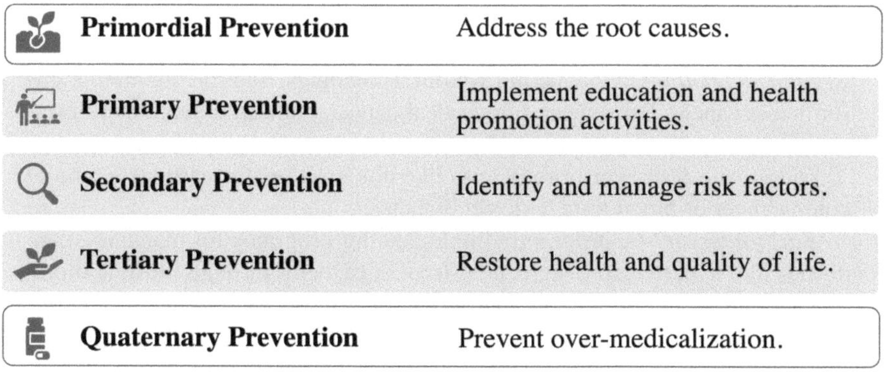

Fig. 4.2 The five level preventative medicine framework

Box 4.1 Examples of Primary Prevention

Health education and promotion: Informing a population about the benefits of a balanced diet through food guides, making recommendations for regular exercise, providing recommendations for reducing alcohol intake, programming or support for smoking cessation, education about safe sex practices, needle exchange programs (Calvo et al., 2021), and micronutrient supplementation programs.

Immunisations: Administering vaccines to protect against infectious diseases such as measles, polo, or influenza or HPV (Maier et al., 2015).

Environmental interventions: Implementing measures to ensure a clean water supply, proper sanitation, and reducing exposure to environmental toxins.

Legislation and policies: Implementing laws and regulations that promote public health, such as seatbelt laws or restrictions on prescription drugs and tobacco advertising.

occurrence of particular diseases or health events in specific regions or populations (AbdulRaheem, 2023; Kisling & M Das, 2023).

Secondary Prevention involves early detection and prompt treatment of a risk state to prevent its progression and minimise its impact. Secondary prevention focuses on individuals who are already at risk or have early signs of a disease but may not yet have developed symptoms. The key objectives of secondary prevention are early diagnosis, early intervention, and preventing manifestations of disease. Secondary prevention strategies are aimed at early stages of disease/injury or illness when it is more manageable and has a higher chance of successful intervention.

Box 4.2 Examples of Secondary Prevention

Regular screenings: Conducting routine screenings, such as mammograms for breast cancer, Pap smears for cervical cancer, cholesterol and blood pressure checks for cardiovascular diseases, and screens for glaucoma.

Diagnostic tests: Performing tests like colonoscopy for early detection of colon cancer or blood tests to detect diabetes.

Early intervention programs: Implementing programs for managing conditions like hypertension, high cholesterol or diabetes through lifestyle modifications, medication, or other appropriate treatments.

Tertiary prevention focuses on managing and rehabilitating individuals who have been diagnosed and have been impacted by a disease or condition. It aims to prevent complications, reduce disability and improve quality of life. Tertiary prevention strategies are implemented after the disease or condition has developed and may involve long-term interventions. Tertiary prevention aims to improve the overall health outcomes of individuals with established diseases.

Box 4.3 Examples of Tertiary Prevention

Rehabilitation programs: Physical therapy, occupational therapy, or speech therapy to help individuals recover or adapt to their condition, such as occupational and physical therapy for stroke patients or burn patients, cardiac rehabilitation for patients post myocardial infarction, and diabetic foot care.

Medical treatment: Administering medication, surgical procedures, or other medical interventions to manage the disease or condition effectively.

Support groups and counselling: Providing emotional support, education, and counselling to patients and their families to cope with the challenges associated with the disease/illness/injury.

As outlined by Abdul Raheem, the medical model of prevention has evolved over time. Two new tiers have been included (Fig. 4.2: indicated in white) that sit on each side of the original framework. These tiers include primordial and quaternary prevention (AbdulRaheem, 2023).

Primordial prevention considers the systemic root causes of disease and the social determinants of health. Interventions and activities that are part of the primordial prevention level aim to build the infrastructures for a healthful society, such as policies that promote health equity or environments that support healthy living. Examples of primordial prevention include government policies that improve access to healthcare, banning smoking in public spaces, developing strategies that address food insecurity and food deserts, and investing in the development of green spaces in urban settings and walkable communities (Martins et al., 2018; Pandve, 2014).

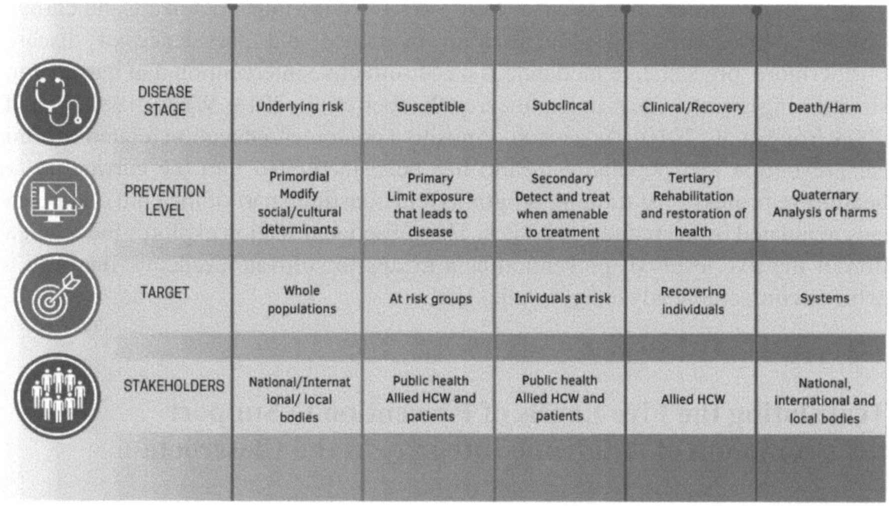

Fig. 4.3 The five levels of prevention framework. *HCW* health care workers

Quaternary prevention includes protective strategies that are specifically focused on reducing the harms related to overmedicalisation (Martins et al., 2018; Pandve, 2014). These could include harms that come from unnecessary treatments of incidental findings, such as radiological incidentalomas (asymptomatic tumours discovered via MRI or CT scans), stewardship over antibiotics to prevent or minimise the impact of multidrug resistant bacteria or reducing the use of antiarrhythmic medications after a myocardial infarction that carry high risk for mortality (Martins et al., 2018). Quaternary prevention involves activities that mitigate or avoid the consequences of unnecessary or excessive intervention. In this tier of prevention, Gofrit et al. (2000) include activities related to quality assurance as well as activities devoted to making improvements in processes, which work together to close the loop of prevention by gathering information about the processes, conducting data analysis, drawing conclusions, and disseminating them to all relevant parties. Reviewing a patient's medications to identify and eliminate any duplicates or unnecessary drugs that could have side effects or using evidence to inform or update guidelines for diagnostics and/or treatments are some examples of quaternary prevention.

The five levels of prevention as they exist today are based on the five stages of disease/illness/injury progression: underlying, susceptible, subclinical, clinical, and recovery/disability/death (Tesser, 2017). Preventive medicine practices, interventions and activities have been grouped into levels that align with these stages (Kisling & M Das, 2023) as illustrated in Fig. 4.3. Each level focuses on a different stage of prevention and contributes to overall healthcare strategies. It is important to note that the levels of prevention are not mutually exclusive and often overlap (Last, 2007). All levels of healthcare systems work together to incorporate intervention strategies to create comprehensive approaches to disease prevention and management.

Preventative medicine has been shown to reduce the impact of acute and chronic disease as measured by reductions in incidence and prevalence of disease. Furthermore, preventative medicine is a cost-effective intervention that has demonstrated long-term savings in healthcare (Webber et al., 2018; WHO, 1986; Woolf, 2008; Yong et al., 2010). As a result of positive outcomes, education related to clinical prevention and population health has been integrated into the curriculum of healthcare professional training programs and constitute part of the core competencies mandated by accreditation bodies. The subsequent section explores the application of the five levels of prevention as a means to cultivate academic integrity to achieve comparable advantages in the HE.

Translating the Five Levels of Prevention to Support the Promotion of Academic Integrity in the Classroom

While academic misconduct is not a disease, illness or injury, it does have similar attributes and patterns to pathology in that misconduct behaviours are influenced by social and cultural determinants, the incidence is monitored to some extent, albeit inconsistently, and incidence and prevalence of misconduct is endemic. Moreover, the goal for interventions is to reduce the risk and prevalence of misconduct and reduce the burden on already constrained systems of education. Similarly, the goal in medicine is to reduce the risk and prevalence of disease to promote a healthy society and reduce the burden on already constrained systems of healthcare. Therefore, it follows that the framework of preventative medicine may be a viable and useful tool for prevention of academic misconduct, empowering both educators and learners by including methods to prevent misconduct as well as tools that equip learners with the knowledge and skills to act with integrity regardless of the circumstance. Each institution, discipline and even classroom will be its own microcosm or ecosystem that will require a bespoke approach in each realm of prevention. Table 4.1 provides a surveying tool that can direct educators in identifying each level.

Using Primary Prevention Strategies to Build a Culture of Academic Integrity

Primary prevention means working with students to develop a shared definition of academic integrity pertinent to the discipline and the learning environment that is relevant and practicable at that time. Primary prevention requires removing the assumptions that academic integrity is one fixed concept and that students come in with an intrinsic understanding of this definition of academic integrity. Rather, acting with integrity is a skill that requires defining and practice. The key elements of primary prevention include: (1) building relationships; (2) developing a

Table 4.1 Translating the five levels of preventative medicine to promotion of academic integrity

Level	Principle activity	Questions to develop practical solutions for preventing misconduct at each level
Primordial	Look for root causes.	What are the systemic social, cultural, and technological determinants that lead to misconduct behaviours?
Primary	Implement education activities.	What can we do to foster a culture of integrity? How do we instil the importance of behaving with integrity in learners? How can we support learners in the development and practice of integrity-based behaviours?
Secondary	Identify and manage risk factors.	What are specific risk factors that lead to misconduct in the classroom? How can curricula and assessments be designed to mitigate these risks?
Tertiary	Restore integrity-based behaviours and relationships.	What methods can be used to help restore integrity behaviours once misconduct has occurred? How can relationships be mended?
Quaternary	Prevent the harms of excessive interventional strategies.	How can we prevent overly rigid approaches to mitigating the risk for misconduct?

discipline-specific, shared definition of what constitutes behaviours related to academic integrity; and (3) providing learners with the opportunity to develop and practise academic integrity within this context.

Building Relationships

Academic integrity and integrity, more broadly, are relational concepts that involve the dynamic interplay between individuals, institutions, and society. To establish a culture of integrity in academia requires fostering trust, accountability, and ethical behaviour within these relational contexts. Integrity begins with the individuals' relationship with themselves. It involves being honest about and true to one's own values, beliefs and principles. From the academic perspective, this means not only refraining from cheating or plagiarism, but also being accountable for one's own work and behaving ethically in the academy. An educator could consider facilitating the development of a learner's intra-personal definition of integrity in the classroom setting by asking the learner to make connections between their values and their learning journey. Integrity is also about how individuals interact with others. Educators can support an inter-personal understanding of academic integrity by prioritising a learning environment that promotes social learning.

Developing Shared and Contextual Definitions of Integrity Behaviours

A truly robust concept of academic integrity that is future proof requires moving away from the compliance model and towards a co-operation model of promoting integrity. Part of this process requires a shift in the way the academic community has traditionally conceptualised academic integrity and academic misconduct.

Tensions arise in a globalised system of HE where learners come from heterogenous social and cultural backgrounds, but the system relies on rigid, antiquated definitions of integrity to which each individual student is meant to ascribe and practise without formal education in this space. Furthermore, there is a mismatch between the theory and practice of integrity in HE, that is, the concept of integrity and the actions related to integrity require a commitment to a social contract, however, in practice, learners in HE are educated and assessed as individuals where adherence to social integrity is rendered meaningless. Scollon (1995) succinctly and tangibly described this idea when he stated: "The concept of plagiarism is fully embedded within a social, political and cultural matrix that cannot be meaningfully separated from its interpretation" (Scollon, 1995, p. 23). In other words, integrity is influenced by changing social, cultural, technological, and political circumstances; and actions of integrity cannot be separated from the social/relational underpinnings, which gives meaning to integrity as a concept.

To overcome these challenges, educators in HE must adopt a dynamic and relational model of academic integrity that accounts for nuances in expression between cultures and disciplines and allows for adaptation to social, cultural, and technological change. Once it is accepted that the concept of academic integrity is not fixed and cannot apply to the individual, the next major challenge revolves around facilitating the development of knowledge and skills around a dynamic and relational concept of academic integrity in our learners. The prevailing assumptions and expectations amongst educators are that the student comes into the HE classroom with a concept of integrity that aligns with the instructor's and that the student is prepared to use this knowledge in every classroom experience (Eaton & Christensen Hughes, 2022; McNeill, 2022). However, this is not the case, particularly for international students (Denisova-Schmidt, 2023; Parnther, 2022; Sanni-Anibire et al., 2021). To meet institutional expectations around academic integrity, students need a sound understanding of what academic integrity is and why it is important. The goal is to create alignment between students' perspectives of academic integrity and the instructors' behavioural expectations. Learners need an opportunity to acquire the skills and to put these skills into practice (Bertram Gallant, 2008, 2017; Morris, 2015). Therefore, starting with constructing a concept of what behaving with integrity means, within a specific circumstance and providing students with the ability to practise those behaviours in an appropriate and authentic way, will help students build their personal understanding and meaning of integrity, which they can take with them outside of the classroom setting.

Communicating, Modelling and Practising Integrity Behaviours in the Classroom

At this stage in their academic career, students are still learning academic and professional standards. In many cases, these standards are not explicitly taught but rather learned through experience. Modelling academic and professional standards and integrity in both our classrooms and in our interactions with students can significantly help the learning process. Modelling helps to demonstrate the norms and standards of integrity behaviours within the discipline as there can be discipline specific interpretations of integrity. Modelling expected behaviours also enables learning through precedent. Modelling integrity can be accomplished by showing students how to complete a task or solve a problem and explicitly stating what is expected of them and how they should accomplish the task by using the standard of practice in written and presented materials. Modelling can also be accomplished through using discipline-specific language, citing sources wherever this is applicable, adding citations below all videos, animations or images in presentation slides, providing references for the research, ideas and concepts presented in class, using resources that are rigorously vetted as credible sources and demonstrating how to find and use credible source material in the discipline. Explicitly pointing to how standards and expectations are being modelled can help to avoid modelling becoming part of the hidden curriculum.

Just as all people have diversity in their experience with and understanding of healthy living, in the diverse setting of academia, learners are heterogenous in their experiences with academic norms and expectations (Schneider, n.d.). Therefore, it is essential to openly communicate academic expectations as well as the rationale behind these standards and related policies (Morris, 2015), in addition to connecting this to professional expectations in the discipline. Educators can do this by including an academic integrity statement on the syllabus, discussing it in class, and reminding students of the relevant policies before major assignments. Moreover, providing examples or interactive opportunities for students to explore acceptable and unacceptable behaviours and explaining the consequences of violating academic integrity standards when completing activities and assessments may be helpful in reinforcing this information.

It can be argued that attention paid to primary prevention can have the biggest impact. This is indeed the case in preventative medicine, where helping healthy individuals develop health promoting behaviours considerably reduces the burden of injury, illness and disease, which in turn reduces the burden on the healthcare system. A notable example of the impact of primary prevention focused on approaches that educate a population comes from the remarkable reduction in the twentieth and twenty-first century of lung cancer and cardiovascular disease that followed declines in smoking accomplished through education campaigns (Surgeon General, 2014, 2020). Vaccination programs provide another example in which primary prevention has had extraordinary success in reducing the burden of disease; this includes significant reductions in the incidence of diseases, such as measles, mumps, rubella, smallpox, pertussis, tetanus, and Haemophilus influenzae type b (Maier et al., 2015).

Similarly, primary prevention strategies directed at promoting integrity through teaching and learning are where the biggest impact can be made (Bertram Gallant, 2017). In her chapter, *Changing Hearts and Minds*, McNeill (2022) supports the use of "pedagogies of integrity," suggesting that "we need to close the gap between our expectations of academic integrity and students' understanding of those expectations and to make explicit what is often treated as assumed understanding" (p. 487).

McNeill's (2022) project provides evidence that integrating efforts of teaching integrity within the curriculum is, in fact, efficacious in reducing the incidence of misconduct, specifically in the form of accidental misconduct. Benson and Enstroem (2023) report that implementation of university-wide core courses and the establishment of common knowledge about academic integrity, reduced the incidence of academic misconduct in the School of Business, having a sustained effect up to 2 years after the intervention was applied.

If primary prevention works to support learners with understanding and practising academic integrity, then emphasising teaching and learning approaches that enable students to build and practise integrity behaviours in a social learning environment will reduce the incidence and prevalence of academic misconduct. It is important to highlight that just as health is not static, integrity is also not static. Through primary prevention and facilitation of this educational process, educators can empower learners with the skills to continually assess their actions and decisions in light of ethical principles, considering the impact on themselves and others through changing circumstances.

Using Secondary Prevention Strategies as Protective Measures to Reduce the Risk of Academic Misconduct

From the healthcare perspective, secondary prevention is about mitigating risk. The team of allied health workers look for opportunities to identify people at risk for a specific disease, illness or injury at a stage where interventions are possible and efficacious, and/or where health promoting activities would be beneficial. In secondary prevention, the patient is at the centre, and interventions focus on how to support the patient to achieve health, whether it is through modification of risk factors or enhancing protective factors. As an example, the reduction in mortality caused by coronary heart disease in the last five decades can, in part, be attributed to preventive medications applied in risk states, such as the identification and treatment of hypertension (Bundy et al., 2017; Carey et al., 2021; Ettehad et al., 2016). Another example is the reduction in breast cancer mortality rates that has been associated with mammography screening (Gelmon & Olivotto, 2002; Marmot et al., 2013; Nelson et al., 2016).

Towards the prevention of academic misconduct, secondary prevention refers to identifying the underlying factors or conditions that lead students to commit misconduct within an academic setting. Instead of simply addressing the surface-level symptoms of misconduct, such as cheating or plagiarism, investigating risk factors involves examining challenges that may contribute to or enable such behaviours. This could include factors, such as academic pressure, lack of understanding of ethical standards, inadequate support systems, or attitudes towards assessments and activities. Identifying challenges students may face that can lead to misconduct behaviours and subsequently designing assessments that engage students and teach them how to avoid these challenges constitute secondary prevention approaches (Bertram Gallant, 2008; Goldman et al., 2022; Kalra, 2019).

Specific risk factors that lead to academic misconduct have been extensively reviewed (Bertram Gallant et al., 2015; Rundle et al., 2019, 2023; Yu et al., 2017). One major risk factor leading to misconduct is related to lack of student engagement. The research suggests that when students are engaged, they are less likely to cheat and more likely to exhibit integrity behaviours. In one study, Bretag et al. (2019) support this, demonstrating that students are more likely to practice unethical academic behaviours when there is "dissatisfaction with the teaching and learning environment" (p. 1846). Furthermore, Coetzee et al. (2022) demonstrate that a sense of belonging, motivation, and hope are all protective against misconduct behaviours. If low student engagement is a risk, then secondary prevention strategies would include intentionally designing in class activities and assessments that create an inclusive and supportive learning environment. Academic misconduct is almost always recognised at the level of assessments. Weak assessment design contributes to a lack of clear expectations, which can cause anxieties for students that lead to learners commit misconduct (Schneider, n.d.). The goal of secondary prevention is to reduce the opportunity or incentive for academic misconduct and promote academic integrity through revisioning of the structure, format, or instruction related to an assessment.

Outcome-based Teaching and Learning (OBTL) is a strategy for designing curriculum elements that starts with identifying what learners should know and/or be able to do at the end of a course or program (Spady & Uy, n.d.; Biggs et al., 2022; Wiggins & McTighe, 2005). These outcomes also specify how students can demonstrate their proficiency, which serves as a template for designing assessments in addition to providing learners with clear guidelines of what is expected of them. While OBTL has been used for decades in some disciplines, it remains underutilised in other areas and, in particular, STEM fields, where incidence and prevalence of academic misconduct remains increasingly problematic. Using a process such as OBTL to design curriculum helps guide educators to deeply consider the what, why, and how of an assessment, precisely defining the expectation of the student (what) and the relevance of this task (why) and explaining the process the student must go through to complete this assessment successfully (how). Sharing the intentions behind the assessment approach with learners provides them with clear and explicit learning goals, shows them the value and relevance of the skills and knowledge they will gain, and gives them clear expectations and standards for their work. This dialogue with learners can reduce the confusion and anxiety related to assessments, increase motivation, engagement and confidence with completing these tasks, and encourage students to ask for help with academic challenges. Under these conditions, learners would theoretically be better equipped to complete course assessments successfully and, therefore, less likely to resort to actions that would be considered as academic misconduct (Bertram Gallant, 2017; Kalra, 2019).

Scaffolding and Sequencing: Scaffolding is the process of breaking down a larger assignment into smaller assignments that focus on component skills or types of knowledge (Ambrose et al., 2010). Encouraging learners to construct the foundational elements of a final product motivates them to complete their own work, as smaller tasks may seem less daunting to accomplish. Sequencing is the process of arranging the scaffolded assignments into an order that builds towards the larger assignment (Ambrose et al., 2010). Using scaffolding in assessment design also enables educators to support learners through the learning process taking the emphasis of the end product. Scaffolding and sequencing together focuses learners on building skills and knowledge and encourages engagement by creating manageable cognitive tasks and by reducing cognitive load (Beasley, 2014; Goldman et al., 2022; McCabe & Pavela, 2004; Sterngold, 2004). A side effect of this strategy is that it also helps students learn how to break large projects up into workable chunks and to develop time management skills.

Authentic and Renewable Assessments: Frey et al. (2012) explain that authentic assessments are tasks that involve "the student deeply, both in terms of cognitive complexity and intrinsic interest and are meant to develop or evaluate skills and abilities that have value beyond the assessment itself" (p. 14). Renewable assessments are "a form of assessment that adds value to the world" (Wiley, 2015, 2016) outside of the classroom setting. Both authentic and renewable assessments measure whether students can apply the knowledge they have gained in a direct and relevant way; therefore, the learning and the assessment are made meaningful (Frey et al., 2012; Mueller, 2018). Authentic problems or tasks, by nature, often offer multiple solutions. In such scenarios, students employ their theoretical understanding to devise somewhat distinct responses to challenges characterised by complexity and ambiguity, which further contributes to deeper, transformative learning (Alhadeff-Jones, 2012; Mezirow & Taylor, 2011). These features of authentic and renewable assessment contribute to building assessment tasks that promote integrity. The data demonstrates that students are more motivated to complete an assignment if they feel that the assignment holds value and is well aligned with course/knowledge objectives or their personal goals (Wigfield & Eccles, 2000). Authentic and renewable assessments encourage academic integrity by employing complex, unique, and creative activities which can make the task itself or the performance of the task more rewarding and meaningful to the learner.

It is critical to highlight that in order for secondary prevention strategies to be most effective, educators require support in the form of time and training. In the Alberta Colleges and Institutes Faculties Association's position statement on academic integrity, Schneider (n.d.) describes how, without proper support, a self-perpetuating cycle of weak assessment design and incidences of misconduct arises. In this cycle, educators take on the initiative to build curriculum elements that are more engaging, including assessments. However, without proper support, most educators are doing this in addition to existing responsibilities and without proper time and training to do the task most effectively. Together, this leads to "weak" or poorly designed assessments that, in turn, lead to lack of engagement for students, potentially triggering more incidences of misconduct and putting further pressures on educators to place more efforts on making engaging assessments without resources (Schneider, n.d.). To stop this cycle, institutions must recognise that maintaining high quality curriculum is a significant task that requires support for not only development and design but also for developing robust research to move interventions from theory to evidence-based practice.

Using Tertiary Prevention Strategies to Promote Future Integrity Behaviours

Tertiary prevention involves managing disease when it occurs and restoring health. For example, in medicine, tertiary prevention may involve rehabilitation programs for patients recovering from surgeries or managing chronic conditions. It also could refer to treating sequelae of disease, such as pain management for cancer patients, or managing complications of disease, such as vascular problems that arise with diabetes mellitus. Tertiary prevention programs not only aim to restore physical health but also provide support for mental and emotional well-being, aiming to reintegrate patients into their daily lives. In both medicine and academic misconduct, tertiary prevention shares a common goal of addressing the consequences of a challenge to health or integrity respectively, after it has occurred. Tertiary prevention in medicine focuses on minimising the impact of diseases or injuries and preventing recurrence through rehabilitation and support services. Similarly, tertiary prevention of academic misconduct would aim to support the learner in maintaining integrity-based behaviours in the future.

Traditionally, when a learner has committed misconduct, some form of reporting and disciplinary action is undertaken in a "quasi-judicial procedure" (Sopcak & Hood, 2022, p. 559). This process is often laid out in policies and procedures that are unique to the department, faculty, and institution. Nonetheless, the approaches tend to focus on a model of compliance, and the measures taken are punitive in nature. Retraining often revolves around academic integrity in general and the process is designed to point students to generic institutional policies and procedures, lacking both discourse and context (Eaton, 2021; Shrolyk & Whitney, 2023). The disciplinary process lacks the social and relational values that are necessary to put integrity and academic integrity into context and then into practice.

Tertiary prevention encourages moving away from compliance and discipline and instead would focus the attention on guiding learners to overcome the challenges that led to misconduct as well as inviting them to reflect on what it means to act with integrity. The student is encouraged to consider intra- and interpersonal impacts of academic misconduct. This approach is formative and supports the development and practice of integrity behaviours in a learning environment which follows a teaching and learning approach to academic integrity. When framed this way, tertiary prevention is about restorative practice.

In recent years, the application of restorative justice to cases of misconduct has gained some traction (Abramson & Rucker, 2023; Kara & MacAlister, 2010; Mullen, 2021; Sopcak & Hood, 2022). Restorative justice emphasises accountability while, at the same time, encouraging the community to come together to support the learner to reflect on the impact of misconduct, repair the harm, and relearn the path to integrity. Restorative justice gives learners an opportunity to gain an understanding of the consequences of their actions. The process may include mediation sessions between students, educators, and administrators, where the focus is on understanding the underlying reasons for the misconduct and collaboratively developing solutions to prevent recurrence.

According to Wachtel (2013), restorative justice falls under the umbrella of a broader range of actions known as restorative practice. With a focus on social connections, restorative practice seeks to achieve several goals: building a strong social fabric, enhancing community resources and social capital, and reducing crime and antisocial behaviour. Additionally, restorative practice works to repair damage and restore relationships in the aftermath of conflict (Costello et al., 2009; Wachtel, 2013). Karp (2019) defines four underlying principles to the application of restorative practice in the promotion of integrity: (1) Inclusive decision making; (2) Active accountability; (3) Repairing harm; and (4) Rebuilding trust (Karp, 2019). Restorative practice can also touch on primary and secondary levels of prevention as it includes the use of informal and formal processes that *precede* wrongdoing, those that proactively build relationships and a sense of community to prevent conflict and misconduct.

Rather than solely focusing on punishment, restorative practices emphasise relational thinking and a sense of accountability within the academic community. This approach acknowledges that instances of misconduct can be acute and symptomatic of deeper issues and seeks to address the underlying causes rather than simply applying punitive measures. Furthermore, restorative practice enables a customised approach to managing instances of misconduct, which takes into consideration nuances of incidents, such as intentionality, so as to not treat deliberate behaviours in the same way that one would treat unintentional violations. This is important, as academic misconduct falls on a spectrum and overt behaviours of misconduct, such as engaging in contract cheating, require different methods of restorative practice than those that would be effective when a student has unintentionally plagiarised. Restorative practices that support re-establishment of academic integrity involve a shift towards more inclusive and collaborative approaches to address misconduct.

Research suggests that implementing restorative practices can lead to more positive outcomes, such as increased trust, improved communication and a greater sense of responsibility among students and faculty (Sopcak & Hood, 2022). For example, the Georgian College Language Help Centre took a restorative approach to manage cases of plagiarism. Students who were identified as potentially at risk of committing or had committed academic misconduct were referred by faculty to "academic honesty" training sessions. Interestingly, the study found that when the students were able to resubmit their papers, they were more motivated to complete academic integrity training and learn the academic writing strategies and rules. The researchers also demonstrated that students who participated in training sessions were less likely to repeat the offence (Shrolyk & Whitney, 2023). Orr and Orr (2021) provide another example of an undisclosed university that made significant revisions to how it approaches academic misconduct. This institution implemented a restorative practice that uses student-led meetings to resolve cases of alleged academic misconduct rather than case adjudication by hearing panels. The researchers demonstrated that the peer-to-peer approach promoted better understanding of expectations around academic conduct and gave students a stronger sense of inclusion (Orr & Orr, 2021).

While there is evidence of its potential benefits, research on restorative justice and academic misconduct is still developing, and more studies are needed to fully understand its long-term impact. Despite the infancy of this approach, it is clear that using restorative justice/practices in managing cases of academic misconduct is in line with the concept of tertiary prevention. In tertiary prevention, the priority is to address the immediate needs of the individual, providing individuals with the tools to understand and take responsibility and be accountable for their own actions, while also supporting rehabilitation of integrity and restoring relationships within the institution.

Roles for Primordial and Quaternary Prevention in Promoting Academic Integrity

Primordial Prevention of academic misconduct involves examining the broader systemic factors that contribute to dishonest behaviours in academia. This tier of prevention involves interrogating the social, cultural, and technological determinants of academic misconduct that extend beyond the confines of the classroom, and the actions of individual students or educators. I argue that these determinants act as the root cause and encompass various aspects of the academic environment and its underlying structures. In my interpretation, an example of a root factor is the focus of academic assessment on the individual over the group. Academic culture often prioritises individual achievement over collaboration and collective success. This emphasis may inadvertently foster a competitive atmosphere where students feel pressured to prioritise personal success at the expense of academic integrity. Relatedly, the highly competitive nature of academics, with high stakes attached to academic achievements, such as grades, publications, and funding opportunities, may incentivise individuals to resort to unethical means, such as cheating or plagiarism, to gain a competitive edge. Finally, I also believe that there is an overreliance on grades as metrics of success, leading to an overemphasis on academic performance as the primary measure of achievement. This narrow focus on grades may overshadow other valuable aspects of education, such as critical thinking, creativity, and ethical conduct.

It could be argued that to address the challenges outlined above, interventions related to primordial prevention start with a much broader critical evaluation of the relevance and efficacy of the current system of education to support the needs of modern society. From a historical perspective, universities served as the only conduit through which knowledge and practical skills were shared with a limited and privileged few. With exponential increases in knowledge and advancements in technologies, the educational landscape is very different from 100 and even 10 years ago, but our institutions of HE and our pedagogies harbour vestiges from their origins thousands of years ago (Duderstadt et al., 2002). It may then follow that academic misconduct behaviours occur as a result of this mismatch (Townley & Parsell, 2005). Karthik Krishnan (2020) calls for education transformation to improve the relevance of education programming which will have a direct effect on expressions of academic integrity (Krishnan, 2020). This type of work can be considered interventions that fit into the level of primordial prevention. Krishnan (2020) suggests that there are three major pillars to focus on with respect to transformation: access, equity and quality/impact. Particularly compelling in relation to academic integrity are the improvements in quality and student success. Improving education by focusing on outcomes, ensuring learners are equipped with the knowledge and skills that would make them job-ready, responsible citizens in this information rich, digital world will improve the quality of education and student outcomes (Krishnan, 2020) while also contributing to a culture that cultivates integrity behaviours. Addressing these systemic determinants of academic misconduct requires interventions that go beyond the classroom and individual disciplinary actions. It involves challenging entrenched norms and practices within the academic community, promoting a culture of collaboration and ethical responsibility, and re-evaluating the metrics used to measure academic success. By interrogating and addressing these broader systemic factors, primordial prevention efforts can help create a more supportive, equitable, and integrity-focused academic environment for all.

Quaternary Prevention in medicine is a relatively new concept in prevention, coined by Jamoulle in 2015. Quaternary prevention is a protective strategy that aims to reduce excessive medicalisation and resulting harm by avoiding overdiagnosis and minimising unnecessary treatments (Jamoulle, 2015; Martins et al., 2018). According to Tesser (2017), "quaternary prevention induces critical evaluation and reaction in physicians, health professionals and professionals and managers of health systems about themselves and their activity in an operational manner, including the questioning of their technical and ethical limits and the recognition of influences that affect decisions and preventive behaviours. It points to the construction of good practices, against cultural, technical and institutional tendencies, which are sometimes harmful to individual and collective health" (Tesser, 2017, p. 3).

On the surface, it may appear that this tier of prevention is not clearly translatable to prevention of academic misconduct and promotion of academic integrity. However, it can be argued that excessive prophylaxis against, or over-treatment of academic misconduct can propagate a cheating culture and a moral panic about academic dishonesty (Bertram Gallant, 2008; Callahan, 2007). As McNeill (2022) states and, as was apparent during the pivot to online learning in 2020, the discourse about academic integrity is based on "deficiency and distrust, focusing on detection rather than education, and perpetuating assumptions about who commits misconduct" (p. 488). Consider, for example, the controversies around the use of proctoring software during online assessments or the initial antagonism to the release of openly accessible generative artificial intelligence tools. There are consequences to the reactionary implementation of these types of preventative approaches in terms of impacts on privacy and security for the former and impeding learning, creativity and innovation, for the latter (Bertram Gallant, 2008). More importantly, it can be argued that overtreatment may even lead to learners being overly fearful of inadvertently committing academic misconduct, or even distrust in the education system as it has in the healthcare system.

The potential consequences of "overmedicalisation" in prevention of academic misconduct include the propagation of a fear of cheating culture (Deters et al., 2020) and the exacerbation of systemic biases (Denisova-Schmidt, 2023; McNeill, 2022; Sanni-Anibire et al., 2021). These are harms related to an over-reliance on traditional, rigid definitions of what constitutes integrity and punitive measures or excessive enforcement of rules and regulations. "Overmedicalisation" could lead to a narrow emphasis on discipline or even strict applications of secondary prevention strategies, rather than proactive measures to promote a culture of integrity and ethical behaviour among students and educators.

Implementing quaternary prevention approaches safeguards against the risk of inadvertently exacerbating issues while attempting to prevent misconduct. Quaternary prevention looks to identify situations where academic institutions focus too heavily on finding, reporting, and punishing instances of academic misconduct without adequately addressing the underlying issues that contribute to such behaviour. Actions that constitute quaternary prevention reduce the risk of the system overreaching to the extent where well-intentioned measures designed to deter misconduct inadvertently result in further harm. This approach involves careful consideration of the potential impacts of preventive measures to avoid causing harm while striving to uphold integrity within academic environments. Furthermore, strategies related to quaternary prevention would include continuous monitoring and refinement of policies so that institutions can ensure that their efforts to promote academic integrity remain relevant, effective, and ethically sound.

Lessons Learned from Preventative Medicine and the Necessity of a Multi-stakeholder Approach

Although there is substantial evidence and professional recognition that the application of five levels of preventative medicine is both cost-effective and impactful (Health Canada, n.d.), implementation of preventative practices remains low, and healthcare resources are still mainly used to manage disease, injury, and illness once it has occurred (AbdulRaheem, 2023). This is an important observation to understand if translation of this prevention framework to academic integrity promotion is to be successful. Several studies have identified the key factors that relate to the poor uptake of the preventative medicine framework (AbdulRaheem, 2023; Kisling & M Das, 2023; Levine et al., 2019). These factors and how they translate to the academic context are described in Table 4.2.

Levine et al. (2019) performed interviews with healthcare experts to ascertain which of these challenges plays the most significant role for the lack of implementation of preventative medicine at the ground level. Their results indicate that the "underutilisation of preventive services is largely the result of an implementation

Table 4.2 Factors that relate to the poor uptake of the preventative medicine framework and how this translates to the academic context

Healthcare context	Academic context
Getting buy-in from healthcare practitioners and patients about the efficacy and benefits of prevention can be challenging.	Getting buy-in from educators, administrators and students about the efficacy and benefits of prevention can be challenging.
Limited awareness and knowledge about the importance of preventive measures among healthcare providers can hinder their adoption.	Limited awareness and knowledge about the importance of preventive measures among educators and administrators can hinder their adoption.
Maintaining currency in the area of prevention can be challenging for physicians and allied healthcare professionals.	Maintaining currency in the area of prevention can be challenging for educators.
Healthcare providers often face time constraints and heavy workloads, which may limit their ability to prioritise preventive care.	Educators face time constraints and heavy workloads, which may limit their ability to prioritise preventive strategies in the classroom.
Ensuring all patients have access to preventative measures and understand the benefits and risks for these measures can be difficult.	It may be difficult to ensure preventative measures, as well as measures to support the development of integrity-based behaviours are equitable and accessible to all students.
Directing resources for preventive services is an effective way to support long-term gain of healthy life years in a population and reduce overall burden on the healthcare system, however, this may be at the expense of supporting critical or acute aspects of healthcare.	Ensuring educators are adequately supported in activities that promote academic integrity and reduce academic misconduct may require redirecting resources away from other activities such as finding, reporting and managing misconduct cases.
The prevention paradox (Rose, 1981) suggests that focusing preventative measures on high risk individuals is less impactful than targeting preventative interventions at low risk individuals as there are a greater number of the latter in the population (Raza et al., 2018; Rose, 1981).	Focusing preventative measures on students at highest risk for misconduct behaviours would be less impactful than putting into practice measures that support all students in developing and practicing integrity-based behaviours.

gap rather than an information gap" (Levine et al., 2019, p. 3). While interviewees were aware of the benefits of preventative medicine, they also highlighted that allied health professionals are compensated for treatment over prevention. Furthermore, participants suggested that changing the fee model to prioritise prevention would be unlikely to improve uptake without increasing systemic supports and resources for implementing preventative medicine. These resources include "organisational leadership, shifts in institutional culture, team-based care, systems of care that accommodate preventive services, and willingness of patients to seek out and engage in preventive care" (Levine et al., 2019, p. 3).

Many of these challenges and obstacles will sound familiar to those working in HE. To effectively address these barriers, AbdulRaheem (2023) suggests a comprehensive and collaborative approach, advocating for strategies like raising awareness, improving access to preventive services, enhancing care coordination and fostering a shift in attitudes and beliefs among healthcare providers and patients.

Examining the question of whose responsibility it is to implement this work, it becomes clear that success for any system of prevention, whether in healthcare or education, lies in fostering strong relationships among all stakeholders involved. Using the healthcare model to illustrate the importance of a multi-stakeholder approach, healthcare professionals, policy makers, organisations and communities have a distinct role to play in integrating prevention in healthcare practice (AbdulRaheem, 2023; Fowler et al., 2020; Kisling & M Das, 2023; White, 2015). Multiple stakeholders collaborate at different levels to promote health and prevent disease. At the local or micro level, responsibility lies with individuals and their allied healthcare teams. Moving up to the meso level, hospitals and municipal organisations are vital contributors to prevention activities and health promotion. At the macro level, governing bodies and institutions like the WHO play crucial roles. These responsibilities are placed within the five levels of prevention as illustrated in Fig. 4.3.

Similarly, prevention of academic misconduct involves a multistakeholder approach. As Kenny and Eaton (2022) highlight, academic integrity transcends student conduct; it encompasses teaching and learning, ethical assessment practices, and decision-making across various roles within the educational institution. Thus, it becomes evident that academic integrity is not solely the responsibility of students but extends to everyone within the "learning ecosystem" (Çelik & Razı, 2023; Eaton & Christensen Hughes, 2022; Whitley & Keith-Spiegel, 2001). In a collaborative approach to academic integrity, students, educators, staff and administrators each hold responsibilities within the learning organisation. This shift towards a shared responsibility model emphasises education and prevention over punitive measures after misconduct occurs. By equipping students with the necessary skills and ethical decision-making capabilities, stakeholders can proactively address academic misconduct. The translation of the five levels of prevention to the promotion of academic integrity is summarised in Fig. 4.4. This figure includes the multiple stakeholders responsible at each tier. At the micro level, educators and students collaborate to create a rich learning environment where educators facilitate the acquisition of skills about what it means to act with integrity, model and demonstrate these characteristics, outline clear expectations and design fair and relevant assessments, while students commit to honestly demonstrate their learning. An important aspect to prevention involves viewing students as partners rather than rule followers. This paradigm shift emphasises cooperation over compliance and empowers students to actively participate in promoting integrity behaviours. At the meso level, institutions: (1) create academic integrity policies and expectations, including definitions of misconduct and consequences for violations; (2) offer educational programs and workshops to raise awareness about academic integrity, proper citation practices, ethical behaviour in research and coursework; and (3) provide training for faculty and staff on recognising and addressing instances of

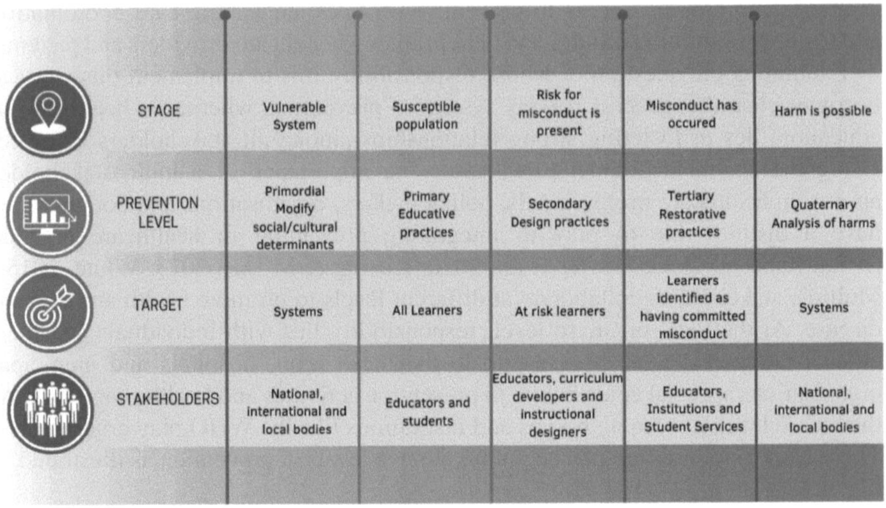

Fig. 4.4 Translating the five levels of prevention to support the promotion of academic integrity in the classroom

academic misconduct, as well as methods for fostering a culture of integrity. Institutions also build the programming and infrastructure for students who may be struggling academically, including tutoring, counselling and academic advising, to reduce the likelihood of resorting to misconduct due to academic pressure. Finally, at the macro level, national and international governing bodies should work to establish frameworks and guidelines to uphold academic integrity and perform quality control while also collaborating with educational institutions to develop standardised protocols, share best practices and implement cross-border initiatives aimed at reducing academic misconduct and promoting academic integrity on a global scale.

Importantly, these stakeholders must not operate in isolation. The key lies in building relationships and fostering cooperation, as integrity holds meaning only when we are interacting with others. Upholding integrity is not an individual endeavour but a collective responsibility requiring collaboration and mutual support. By working in partnerships, stakeholders can adapt and evolve their concepts of integrity to meet the evolving landscapes of education. Thus, fostering strong relationships among stakeholders is paramount in promoting and maintaining academic integrity within educational institutions.

Conclusions

Prevention of academic misconduct and promotion of academic integrity requires a holistic approach that starts with developing a shared understanding of what it means to act with integrity and ending with a critical assessment of whether policies and practices designed to reduce the risk for misconduct are ethical. The five levels of preventative medicine provide an evidence-based framework that can be translated to promoting academic integrity in the academy. Primordial prevention demands that we examine systemic causes of misconduct, primary prevention works to foster a culture of integrity, secondary prevention screens for risk factors and supports the development of interventions within the curriculum that can help to mitigate misconduct, tertiary prevention supports the restoration of integrity and quaternary prevention ensures that interventions remain relevant and ethical. All five tiers of prevention work together towards promoting a culture of integrity, reducing the burden on students, educators and institutions of HE, ultimately leading to reduced instances of academic misconduct and enhancing student engagement and learning outcomes.

Acknowledgements This work would not have been possible without the support of Dr. Laurie McNeil, Dr. Sarah Elaine Eaton, Parisa Zitouni, and Dr. Jacquie Brinkman, who contributed their time to scholarly discussions and/or editorial reviews.

References

AbdulRaheem, Y. (2023). Unveiling the significance and challenges of integrating prevention levels in healthcare practice. *Journal of Primary Care & Community Health, 14*, 21501319231186500. https://doi.org/10.1177/21501319231186500

Abramson, A., & Rucker, A. (2023). Building a culture of academic integrity through restorative justice. *Canadian Perspectives on Academic Integrity, 6*(1). https://doi.org/10.11575/cpai.v6i1.76531

Alhadeff-Jones, M. (2012). Transformative learning and the challenges of complexity. In E. W. Taylor, P. Cranton, & Associates (Eds.), *Handbook of transformative learning: Theory, research and practice* (pp. 178–194). Jossey-Bass.

Ambrose, S. A., Bridges, M. W., DiPietro, M., Lovett, M. C., & Norman, M. K. (2010). *How learning works: Seven research-based principles for smart teaching*. John Wiley & Sons.

Baumann, L. C., & Ylinen, A. (2020). Prevention: Primary, secondary, tertiary. In M. D. Gellman (Ed.), *Encyclopedia of behavioral medicine* (pp. 1738–1740). Springer. https://doi.org/10.1007/978-3-030-39903-0_135

Beasley, E. M. (2014). Students reported for cheating explain what they think would have stopped them. *Ethics & Behavior, 24*(3), 229–252. https://doi.org/10.1080/10508422.2013.845533

Benson, L., & Enstroem, R. (2023). A model for preventing academic misconduct: Evidence from a large-scale intervention. *International Journal for Educational Integrity, 19*(25), 1–17. https://doi.org/10.1007/s40979-023-00147-y

Bentzen, N. (2003). *Dictionary of General Practice Primary Health Care Classification Consortium - WICC WONCA*. http://www.ph3c.org/4daction/w3_CatVisu/en/wonca-dictionary-of-general/family-practice.-2003.html?wDocID=92

Bertram Gallant, T. (2008). *Academic integrity in the 21st century: A teaching and learning imperative* (ASHE higher education report, Vol. 33, No 5). Jossey-Bass. https://doi.org/10.1002/aehe.3305
Bertram Gallant, T. (2017). Academic integrity as a teaching & learning issue: From theory to practice. *Theory into Practice, 56*(2), 88–94. https://doi.org/10.1080/00405841.2017.1308173
Bertram Gallant, T., Binkin, N., & Donohue, M. (2015). Students at risk for being reported for cheating. *Journal of Academic Ethics, 13*(3), 217–228. https://doi.org/10.1007/s10805-015-9235-5
Biggs, J., Tang, C., & Kennedy, G. (2022). *Teaching for quality learning at university* (5th ed.). McGraw-Hill Education.
Bretag, T. (2016). *Handbook of academic integrity* (1st ed.). Springer.
Bretag, T., Harper, R., Burton, M., Ellis, C., Newton, P., Rozenberg, P., Saddiqui, S., & van Haeringen, K. (2019). Contract cheating: A survey of Australian university students. *Studies in Higher Education, 44*, 1837. https://doi.org/10.1080/03075079.2018.1462788
Bundy, J. D., Li, C., Stuchlik, P., Bum, X., Kelly, T. N., Mills, K. T., He, H., Chen, J., Whelton, P. K., & He, J. (2017). Systolic blood pressure reduction and risk of cardiovascular disease and mortality: A systematic review and network meta-analysis. *Journal of the American Medical Association Cardiology, 2*, 775–781. https://doi.org/10.1001/jamacardio.2017.1421
Callahan, D. (2007). *The cheating culture: Why more Americans are doing wrong to get ahead*. Houghton Mifflin Harcourt.
Calvo, F., Carbonell, X., Rived, M., & Giralt, C. (2021). Cuando las personas que consumen drogas inyectadas tienen la palabra: Análisis cualitativo de contenido temático sobre la percepción de uso de una aplicación móvil para los programas de intercambio de jeringas [When people who inject drugs speak: Qualitative thematic analysis of the perception of a mobile app for needle exchange programs]. *Adicciones, 33*(3), 217–234. https://doi.org/10.20882/adicciones.1334
Carey, R. M., Wright, J. T., Jr., Taler, S. J., & Whelton, P. K. (2021). Guideline-driven management of hypertension: An evidence-based update. *Circulation Research, 128*(7), 827–846. https://doi.org/10.1161/CIRCRESAHA.121.318083
Çelik, Ö., & Razı, S. (2023). Facilitators and barriers to creating a culture of academic integrity at secondary schools: An exploratory case study. *International Journal for Educational Integrity, 19*(4), 1–30. https://doi.org/10.1007/s40979-023-00125-4
Coetzee, T., Pryce-Jones, K., Grant, L., & Tindle, R. (2022). Hope moderates the relationship between students' sense of belonging and academic misconduct. *International Journal of Educational Integrity, 18*(1), 28. https://doi.org/10.1007/s40979-022-00121-0
Cohen, L., & Swift, S. (1999). The spectrum of prevention: Developing a comprehensive approach to injury prevention. *Injury Prevention: Journal of the International Society for Child and Adolescent Injury Prevention, 5*, 203–207. https://doi.org/10.1136/ip.5.3.203
Costello, B., Wachtel, J., & Wachtel, T. (2009). *The restorative practices handbook: For teachers, disciplinarians and administrators*. International Institute for Restorative Practices.
Denisova-Schmidt, E. (2023). Academic integrity and international students: An inclusive approach. In S. E. Eaton (Ed.), *Second handbook of academic integrity* (pp. 1187–1200). https://doi.org/10.1007/978-3-031-54144-5_130
Deters, J. R., Paretti, M. C., & Case, J. M. (2020). How implicit assumptions about engineering impacted teaching and learning during COVID-19. *Advances in Engineering Education, 8*(4), 1–5. https://files.eric.ed.gov/fulltext/EJ1287294.pdf
Dowdle, W. R. (1998). The principles of disease elimination and eradication. *Bulletin of the World Health Organization, 76*(2), 22–25.
Duderstadt, J. J., Atkins, D. E., & Van Houweling, D. E. (2002). *Higher education in the digital age: Technology issues and strategies for American colleges and universities*. Greenwood Publishing Group. https://doi.org/10.5040/9798216193616
Eaton, S. E. (2021). *Plagiarism in higher education: Tackling tough topics in academic integrity*. Bloomsbury Publishing. https://doi.org/10.5040/9798400697142
Eaton, S. E. (2024). *Second handbook of academic integrity*. Springer.

Eaton, S. E., & Christensen Hughes, J. (2022). *Academic integrity in Canada: An enduring and essential challenge*. Springer.

Ettehad, D., Emdin, C. A., Kiran, A., Anderson, S. G., Callender, T., Emberson, J., Chalmers, J., Rodgers, A., & Rahimi, K. (2016). Blood pressure lowering for prevention of cardiovascular disease and death: A systematic review and meta-analysis. *The Lancet, 387*(10022), 957–967. https://doi.org/10.1016/S0140-6736(15)01225-8

Fowler, T., Garr, D., Mager, N. D. P., & Stanley, J. (2020). Enhancing primary care and preventive services through Interprofessional practice and education. *Israel Journal of Health Policy Research, 9*(12), 1–5. https://doi.org/10.1186/s13584-020-00371-8

Frey, B. B., Schmitt, V. L., & Allen, J. P. (2012). Defining authentic classroom assessment. *Practical Assessment, Research, and Evaluation, 17*(1), n2. https://doi.org/10.7275/sxbs-0829

Gelmon, K. A., & Olivotto, I. (2002). The mammography screening debate: Time to move on. *The Lancet, 359*(9310), 904–905. https://doi.org/10.1016/S0140-6736(02)08012-1

Gofrit, O. N., Shemer, J., Leibovici, D., Modan, B., & Shapira, S. C. (2000). Quaternary prevention: A new look at an old challenge. *The Israel Medical Association Journal: IMAJ, 2*(7), 498–500.

Goldman, J., Carsons, M. L., & Simonds, J. (2022). It's in the pedagogy: Evidence-based practices to promote academic integrity. In D. Rettinger & T. Bertram (Eds.), *Cheating academic integrity lessons from 30 years of research* (pp. 131–168). John Wiley & Sons.

Green, L. W. (1979). National policy in the promotion of health. *International Journal of Health Education, 17*(3), 91–98. https://doi.org/10.1080/03073289.1979.10805458

Health Canada. (n.d.). *Investing in prevention - The economic perspective* [Government report]. https://www.canada.ca/en/public-health/services/health-promotion/population-health/investing-prevention-economic-perspective-key-findings.html

Institute of Medicine (U.S.). Division of health promotion and disease prevention. (1980). *Perspectives on health promotion and disease prevention in the United States*.

Jamoulle, M. (2015). Prevenção quaternária: Primeiro não causar dano. *Revista Brasileira de Medicina de Família E Comunidade, 10*(35), 1–3. https://doi.org/10.5712/rbmfc10(35)1064

Kalra, J. (2019). *Encouraging academic integrity through intentional assessment design*. Pressbooks. https://pressbooks.bccampus.ca/encourageacademicintegrity/

Kara, F., & MacAlister, D. (2010). Responding to academic dishonesty in universities: A restorative justice approach. *Contemporary Justice Review, 13*(4), 443–453. https://doi.org/10.1080/10282580.2010.517981

Karp, D. R. (2019). *The little book of restorative justice for colleges and universities: Repairing harm and rebuilding trust in response to student misconduct*. Simon and Schuster.

Kenny, N., & Eaton, S. E. (2022). Academic integrity through a SoTL lens and 4M framework: An institutional self-study. In S. E. Eaton & J. Christensen Hughes (Eds.), *Academic integrity in Canada: An enduring and essential challenge*. Springer. https://link.springer.com/book/10.1007/978-3-030-83255-1

Kisling, L. A., & M Das, J. (2023). Prevention strategies. In *StatPearls*. StatPearls Publishing.

Krishnan, K. (2020, April 13). *Our education system is losing relevance. Here's how to unleash its potential*. World Economic Forum. https://www.weforum.org/agenda/2020/04/our-education-system-is-losing-relevance-heres-how-to-update-it/

Last, J. M. (2007). *A dictionary of public health*. Oxford University Press. https://doi.org/10.1093/acref/9780195160901.001.0001

Levine, S., Malone, E., Lekiachvili, A., & Briss, P. (2019). Health care industry insights: Why the use of preventive services is still low. *Preventing Chronic Disease, 16*, E30. https://doi.org/10.5888/pcd16.180625

Macdonald, R., & Carroll, J. (2006). Plagiarism—A complex issue requiring a holistic institutional approach. *Assessment and Evaluation in Higher Education, 31*(2), 233–245. https://doi.org/10.1080/02602930500262536

Maier, C., Maier, T., Neagu, C. E., & Vlădăreanu, R. (2015). Romanian adolescents' knowledge and attitudes towards human papillomavirus infection and prophylactic vaccination. *European Journal of Obstetrics, Gynecology, and Reproductive Biology, 195*, 77–82. https://doi.org/10.1016/j.ejogrb.2015.09.029

Marmot, M. G., Altman, D. G., Cameron, D. A., Dewar, J. A., Thompson, S. G., & Wilcox, M. (2013). The benefits and harms of breast cancer screening: An independent review. *British Journal of Cancer, 108*(11), 2205–2240. https://doi.org/10.1038/bjc.2013.177

Martins, C., Godycki-Cwirko, M., Heleno, B., & Brodersen, J. (2018). Quaternary prevention: Reviewing the concept. *The European Journal of General Practice, 24*(1), 106–111. https://doi.org/10.1080/13814788.2017.1422177

McCabe, D. L., & Pavela, G. (2004). Ten (updated) principles of academic integrity: How faculty can foster student honesty. *Change: The Magazine of Higher Learning, 36*(3), 10–15. https://doi.org/10.1080/00091380409605574

McNeill, L. (2022). Changing "hearts" and "minds": Pedagogical and institutional practices to foster academic integrity. In *Academic integrity in Canada* (pp. 487–503). https://doi.org/10.1007/978-3-030-83255-1_25

Mezirow, J., & Taylor, E. W. (2011). *Transformative learning in practice: Insights from community, workplace, and higher education*. John Wiley & Sons.

Morris, E. J. (2015). Academic integrity: A teaching and learning approach. In T. Bretag (Ed.), *Handbook of academic integrity* (pp. 1037–1053). https://doi.org/10.1007/978-981-287-098-8_11

Mueller, J. (2018). *Authentic assessment toolbox*. http://jfmueller.faculty.noctrl.edu/toolbox/whatisit.htm

Mullen, S. R. (2021). *Efficacy of restorative justice strategies in response to academic dishonesty: Perceptions of high school administrators, teachers and parents*. Doctoral dissertation, Widener University.

Nelson, H. D., Fu, R., Cantor, A., Pappas, M., Daeges, M., & Humphrey, L. (2016). Effectiveness of breast cancer screening: Systematic review and meta-analysis to update the 2009 U.S. preventive services task force recommendation. *Annals of Internal Medicine, 164*(4), 244–255. https://doi.org/10.7326/M15-0969

Orr, J. E., & Orr, K. (2021). Restoring honor and integrity through integrating restorative practices in academic integrity with student leaders. *Journal of Academic Ethics, 21*(1), 55–70. https://doi.org/10.1007/s10805-021-09437-x

Pandve, H. (2014). Changing concept of disease prevention: From primordial to quaternary. *Archives of Medicine and Health Sciences, 2*(2), 254–256. https://doi.org/10.4103/2321-4848.144366

Parnther, C. (2022). International students and academic misconduct: Considering culture, community, and context. *Journal of College and Character, 23*(1), 60–75. https://doi.org/10.1080/2194587X.2021.2017978

Public Health. (2021, November 11). *WHO definition of health*. https://www.publichealth.com.ng/world-health-organizationwho-definition-of-health/who-definition-of-health-2/

Public Health Agency of Canada. (2001). *Population health promotion: An integrated model of population health and health promotion*. Public Health Agency of Canada. https://www.canada.ca/en/public-health/services/health-promotion/population-health/population-health-promotion-integrated-model-population-health-health-promotion.html#toc

Raza, S. A., Salemi, J. L., & Zoorob, R. J. (2018). Historical perspectives on prevention paradox: When the population moves as a whole. *Journal of Family Medicine and Primary Care, 7*(6), 1163–1165. https://doi.org/10.4103/jfmpc.jfmpc_275_18

Rettinger, D. A., & Bertram Gallant, T. (2022). *Cheating academic integrity: Lessons from 30 years of research*. John Wiley & Sons.

Rose, G. (1981). Strategy of prevention: Lessons from cardiovascular disease. *British Medical Journal, 282*(6279), 1847–1851. https://doi.org/10.1136/bmj.282.6279.1847

Rundle, K., Curtis, G. J., & Clare, J. (2019). Why students do not engage in contract cheating. *Frontiers in Psychology, 10*, 2229. https://doi.org/10.3389/fpsyg.2019.02229

Rundle, K., Curtis, G. J., & Clare, J. (2023). Why students do not engage in contract cheating: A closer look. *International Journal for Educational Integrity, 19*(11), 1–21. https://doi.org/10.1007/s40979-023-00132-5

Sanni-Anibire, H., Stoesz, B. M., Gervais, L., & Vogt, L. (2021). International students' knowledge and emotions related to academic integrity at Canadian postsecondary institutions. *International Journal for Educational Integrity, 17*(21), 1–15. https://doi.org/10.1007/s40979-021-00088-4

Schneider, B. (n.d.). *ACIFA's position on academic integrity.* https://www.acifa.ca/_files/ugd/16fcf0_c98c6fb9f9bc4e4880a9f875d754eb4e.pdf

Scollon, R. (1995). Plagiarism and ideology: Identity in intercultural discourse. *Language in Society, 24*(1), 1–28. https://doi.org/10.1017/S0047404500018388

Shrolyk, H., & Whitney, C. (2023). Taking a restorative approach toward academic integrity for L2 Writers. *The Interdisciplinary Journal of Student Success, 2*, 38–47. https://cdspress.ca/wp-content/uploads/2023/07/CALL-Restorative-Approach_Academic-Integrity_Shrolyk.pdf

Sopcak, P., & Hood, K. (2022). Building a culture of restorative practice and restorative responses to academic misconduct. In S. E. Eaton & J. C. Hughes (Eds.), *Academic integrity in Canada* (pp. 553–571). https://doi.org/10.1007/978-3-030-83255-1_29

Spady, W. G., & Uy, F. A. A. (n.d.). *Outcome-based education: Critical issues and answers.* https://eric.ed.gov/?id=ED380910

Stephens, J. M. (2016). Creating cultures of integrity: A multilevel intervention model for promoting academic honesty. In T. Bretag (Ed.), *Handbook of academic integrity* (pp. 995–1007). https://doi.org/10.1007/978-981-287-098-8_13

Stephens, J. M. (2019). Natural and normal, but unethical and evitable: The epidemic of academic dishonesty and how we end it. *Change: The Magazine of Higher Learning, 51*, 8–17. https://doi.org/10.1080/00091383.2019.1618140

Sterngold, A. (2004). Confronting plagiarism: How conventional teaching invites cyber-heating. *Change: The Magazine of Higher Learning, 36*(3), 16–21. http://www.jstor.org/stable/40177968

Surgeon General. (2014). *The health consequences of smoking—50 years of progress: A report of the surgeon general.* Centers for Disease Control and Prevention. https://www.ncbi.nlm.nih.gov/books/NBK294310/

Surgeon General. (2020). *Smoking cessation: A report of the surgeon general.* US Department of Health and Human Services, Public Health Service, Office of the Surgeon General. https://www.hhs.gov/sites/default/files/2020-cessation-sgr-full-report.pdf

Tesser, C. D. (2017). Por que é importante a prevenção quaternária na prevenção? [Why is quaternary prevention important in prevention?]. *Revista de Saude Publica, 51*, 116. https://doi.org/10.11606/S1518-8787.2017051000041

Townley, C., & Parsell, M. (2005). Technology and academic virtue: Student plagiarism through the looking glass. *Ethics and Information Technology, 6*(4), 271–277. https://doi.org/10.1007/s10676-005-5606-8

Wachtel, T. (2013). *Defining restorative.* International Institute for Restorative Practice.

Webber, L., Chalkidou, K., Morrow, S., Ferguson, B., & McPherson, K. (2018). What are the best societal investments for improving people's health? *British Medical Journal, 30*, 362. https://doi.org/10.1136/bmj.k3377

White, F. (2015). Primary health care and public health: Foundations of universal health systems. *Medical Principles and Practice: International Journal of the Kuwait University, Health Science Centre, 24*(2), 103–116. https://doi.org/10.1159/000370197

Whitley, B. E., Jr., & Keith-Spiegel, P. (2001). *Academic dishonesty: An educator's guide.* Psychology Press. https://doi.org/10.4324/9781410604279

Wigfield, A., & Eccles, J. S. (2000). Expectancy-value theory of achievement motivation. *Contemporary Educational Psychology, 25*(1), 68–81. https://doi.org/10.1006/ceps.1999.1015

Wiggins, G., & McTighe, J. (2005). *Understanding by design* (2nd ed.). Association for Supervision and Curriculum Development (ASCD).

Wiley, D. (2015, August 3). An obstacle to the ubiquitous adoption of OER in US higher education [Blog post]. *Open Content.* https://opencontent.org/blog/archives/3941

Wiley, D. (2016). Toward renewable assessment [Blog post]. *Open Content.* https://opencontent.org/blog/archives/4691

Woolf, S. H. (2008). The power of prevention and what it requires. *Journal of the American Medical Association, 299*(20), 2437–2439. https://doi.org/10.1001/jama.299.20.2437

World Health Organization. (1986). *Ottawa charter for health promotion.* http://www.who.int/healthpromotion/conferences/previous/ottawa/en/

Yong, P. L., Saunders, R. S., & Olsen, L. (2010). Institute of Medicine (US) roundtable on evidence-based medicine. In *The healthcare imperative: Lowering costs and improving outcomes: Workshop series summary*. National Academies Press.

Yongu, W. T. (2022). A critical review of levels of disease development and prevention in epidemiology. *International Journal of Human Kinetics, Health and Education, 6*(2) https://journals.aphriapub.com/index.php/IJoHKHE/article/view/1490

Yu, H., Glanzer, P. L., & Johnson, B. (2017). *Why students cheat: A conceptual framework of personal, contextual, and situational factors*. IGI Global.

Chapter 5
A Detailed Guide to Structural Topic Modelling in Educational Research: Insights from a Study on the Ethical Use of AI in Academia

Michael Weiss, Sabiha Mumtaz, and Jamie J. Carmichael

Abstract This chapter presents an innovative approach to analysing survey data on the ethical use of AI tools, such as ChatGPT, in academia. Using a mix of closed- and open-ended survey questions, we collected diverse perspectives from graduate students from different cultural backgrounds. These students were enrolled in graduate management programmes on two campuses of a top international university. Closed-ended questions are efficient for data collection and analysis but may limit exploration, while open-ended questions provide deeper insights but require more resources for analysis. To analyse the open-ended responses more efficiently, we used Structural Topic Modelling (STM), a machine learning technique that identifies themes in open-ended responses and assesses the influence of covariates, such as gender and culture, on those themes. We also used ChatGPT to classify cultural clusters according to a taxonomy from existing research. The main contribution of this chapter is methodological. It provides detailed guidance on the methods used to help other researchers replicate the approach. In addition, this chapter also offers valuable insights into the ethical considerations of using AI tools in education.

Keywords AI tools in academia · Ethical considerations · ChatGPT · Structural Topic Modelling · Survey data analysis · Cultural diversity

M. Weiss (✉) · J. J. Carmichael
Carleton University, Ottawa, ON, Canada
e-mail: MichaelWeiss@cunet.carleton.ca; JamieCarmichael@cunet.carleton.ca

S. Mumtaz
University of Wollongong in Dubai, Dubai, UAE
e-mail: SabihaMumtaz@uowdubai.ac.ae

I. Glendinning et al. (eds.), *Ethics and Integrity in Education (Research)*, Ethics and Integrity in Educational Contexts 9,
https://doi.org/10.1007/978-3-031-98157-9_5

Introduction

The use of artificial intelligence (AI) in academia is raising ethical questions. As AI tools like ChatGPT become more widely used in education, we need to assess their impact on student learning, what we consider to be ethical behaviour and institutional policies. However, understanding student perspectives on these issues requires scalable methodologies that effectively capture quantitative trends and qualitative nuances. We address this challenge by using a semi-automated mixed methods approach to analyse survey data.

Mixed methods research combines quantitative and qualitative approaches to answer research questions. While traditional quantitative surveys capture broad trends, they often lack depth (Creswell & Plano Clark, 2017). Qualitative methods provide more nuanced insights, but are hard to analyse at scale (Roberts et al., 2014). Previous studies have demonstrated the potential of mixed methods for academic integrity research (Newton, 2024), but the challenge of efficiently analysing large numbers of open-ended survey responses remains unresolved.

Closed-ended survey questions offer advantages: they are easier to collect and analyse than qualitative data in the form of open-ended, free-text questions. However, closed-ended questions also restrict what can be measured to predefined categories (Schmiedel et al., 2019), thereby limiting what can be discovered. Open-ended questions provide direct insight into respondents' thinking and, unlike closed-ended questions, do not "cue" respondents to expected categories of answers (Roberts et al., 2014). They allow for inductive theorising, which can lead to the discovery of novel constructs and relationships (Hannigan et al., 2019). Despite their value, analysing open-ended responses can be time-consuming. Systematic approaches, supported by computational tools, are needed to deal with the complexity of unstructured textual data (Schmiedel et al., 2019).

To address this gap, we apply Structural Topic Modelling (STM), a machine learning technique that identifies themes in open-ended responses and assesses the influence of covariates, such as gender and culture, on those themes (Roberts et al., 2014). STM builds on traditional topic modelling by allowing researchers to examine how different factors, such as gender, influence topic prevalence within a dataset. It can complement traditional qualitative methods by providing a systematic way to analyse large text datasets, potentially revealing patterns that might be missed in smaller-scale studies. Its unbiased discovery process reveals emergent themes, offering a comprehensive understanding. Furthermore, STM's combination of quantitative and qualitative insights deepens the analysis, providing a richer understanding of the issues. By using this evidence-based approach, academic integrity researchers can gain critical insights into the root causes of academic dishonesty and suggest more relevant and targeted policies to address these challenges effectively.

Computational methods for analysing text data have become increasingly popular in many disciplines due to the large amount of text data available (Schmiedel et al., 2019). Text data is unstructured and can be difficult to analyse using traditional, deductive methods that require predetermined categories or scales. Even when categories are known, classifying documents using conventional supervised machine learning models requires large amounts of human-annotated training data (Roberts et al., 2014). Thus, computational methods have been proposed to scale our ability to analyse text data while still allowing for human interaction in the process.

Topic modelling (Blei et al., 2003) is an example of a computational method that can be used to discover topics or themes from text data in an inductive manner, without the need for predefined categories. Topic modelling can help researchers look at their data in new and perhaps surprising ways, sometimes suggesting topics that are not immediately obvious to human readers (Nelson, 2020). It can be viewed as an abductive approach that involves observing the data and generating plausible hypotheses based on the patterns and topics identified in the data. Here, we follow Schvaneveldt and Cohen's (2010) distinction between deductive, inductive and abductive approaches to research. Whereas in an inductive approach, observations are made to confirm a hypothesis, in an abductive approach, hypotheses offer plausible explanations of the observations. It is exploratory rather than confirmatory.

STM extends classical topic modelling by including document metadata (covariates), such as gender, political affiliation, or treatment, in the estimation of document-topic and topic-word distributions (Roberts et al., 2014). This allows researchers to estimate the effect of covariates on topic prevalence (i.e., how often a topic is discussed in a corpus). At the heart of this estimation process is a regression model in which topic prevalence is the independent variable, and covariates are the dependent variables (Roberts et al., 2014). STM was originally developed to analyse surveys with a mix of open- and closed-ended questions. Roberts et al. (2019) provide an R package (*stm*) with high-level functions that researchers can use to estimate the effects of covariates on topics and interpret the results. The package also includes functions for visualising the results.

Schmiedel et al. (2019) define a protocol for conducting studies that use structural topic modelling to analyse textual datasets such as those obtained from interviews or open-ended survey questions. The protocol includes steps for inducing topics from textual data, estimating the effect of covariates on topics, and interpreting the results. Weiss (2021) describes common patterns (or practices) that researchers can follow to interpret topics and to assign labels to them. For example, the author recommends selecting a number of topics that ensures they are neither too general nor too narrow, but optimally distinct.

We also use ChatGPT to annotate survey responses. Large Language Models (LLMs) such as ChatGPT and GPT-4 are a recent addition to the quiver of tools for analysing text data. They can be used to generate summaries, answer questions, and classify text into categories or codes (De Paoli, 2023; Heseltine & Clemm von Hohenberg, 2024). The paper by Heseltine and Clemm von Hohenberg (2024) is an example of using LLMs for annotating text. They advocate for a "hybrid" approach to coding with human oversight.

De Paoli (2023) presents an approach to conduct thematic analysis of semi-structured interviews using an LLM. The author shows that using an LLM to perform an inductive thematic analysis is viable and produces reliable results. The

LLM can help generate initial codes, identify themes, and name themes using an inductive rather than a deductive approach. Like Heseltine and Clemm von Hohenberg (2024), De Paoli (2023) recommends a model in which humans and AI collaborate. Our approach is aligned with this vision.

The main contribution of this chapter is methodological. We show how to use Structural Topic Modelling (STM) to analyse survey data containing a mix of closed- and open-ended questions and show how to incorporate covariates such as gender and culture into the analysis. This approach allows a researcher to identify topics or themes in open-ended survey responses and to estimate the impact of covariates on those topics. While they have been used in other fields, the use of topic modelling and structural topic modelling, in particular, is new to the academic integrity literature (Carmichael & Eaton, 2023).

The remainder of the paper is organised as follows. Section 2 describes the method we used to analyse the survey data, which includes structural topic modelling and data preprocessing using ChatGPT. Section 2.2 presents the application of the method to a case study on the ethical use of AI tools, such as ChatGPT, in academia. Section 3 discusses the findings and reflects on new insights that we can gain from them. Section 4 concludes the paper.

Method

Our method for analysing the survey data includes structural topic modelling and data pre-processing using ChatGPT. To create a structural topic model from the survey data, we follow the protocol introduced by Schmiedel et al. (2019). This protocol is supported by the *stm* package in R (Roberts et al., 2019).

Steps

The method consists of the following steps, which are numbered to align with steps in the protocol (e.g., steps 2a and 2b are both data preparation steps).

Step 1: Build a Corpus of Survey Responses

Collect survey data consisting of both closed- and open-ended questions. Determine which text responses will be used as the corpus for the structural topic model, and select covariates whose effect on the text responses you wish to examine. Covariates can be numerical or categorical variables if they are based on responses to closed-ended questions, or they can be derived from responses to open-ended questions by annotating the data. ChatGPT can be used to annotate responses and help to classify and organise the data. Prompt ChatGPT with the classification goal and the codes to use before presenting it with the data to annotate. Including examples or guidelines for handling special cases can ensure consistent and accurate annotations.

Step 2a: Conduct an Exploratory Data Analysis

Conduct an exploratory data analysis on the distribution of the covariates. This step will help you to understand the characteristics of the covariates, e.g., by plotting their distribution. Look out for imbalances in the data or outliers in the responses. These need to be taken into account when you interpret the results.

Step 2b: Preprocess the Corpus

Preprocessing of the corpus includes tokenising the responses, removing common English stopwords (e.g., "the", "is", or "and"), stemming the words (i.e., reducing words to their root form), and converting them to lowercase. Removing stopwords reduces noise in the topics; highly frequent words don't add any meaning to individual topics, if they are included in most topics.

Stemming and converting to lowercase reduce the dimensionality of the dataset and make the results easier to interpret. For example, the words "use", "used", and "using" are all mapped to the common root "use". Therefore, all variants of "use" will occupy only one dimension.

To understand the result of preprocessing, consider an example response:

"To clarify doubt and get basic concepts"

After tokenising, removing stopwords, stemming, and converting the sentence to lowercase, the list of words that remains is:

clarifi, doubt, get, basic, concept

The tokens "clarifi" and "concept" are the stems of "clarify" and "concepts".

Step 3a: Select the Number of Topics

Select a suitable number of topics for the topic model by searching over different numbers of topics and comparing the semantic coherence of the resulting models. Semantic coherence measures the internal coherence of the topics and is maximised when the highest-probability words in a topic frequently occur together (Mimno et al., 2011). Finding the right number of topics is a balance. Too few topics will lead to overly broad, less coherent topics. When there are too many topics, they will split into more fragmented, less coherent, and even duplicate topics (Weiss, 2021).

Step 3b: Label the Topics

Label the topics based on words and documents that are highly associated with each topic. For example, if a topic contains words like “soccer”, “hockey”, “ball”, and “puck”, a good label for the topic would be “sports”. It helps to create word clouds of the most frequent words in each topic. Together with the top documents for a topic, word clouds allow you to “see” the topic as a whole and to assign a label to it. Other metrics can also be computed, such as the most frequent but exclusive words, which are words that distinguish topics. Weiss (2021) provides further heuristics on how to interpret a topic model.

Step 4: Identify the Relationships between Topics and Covariates

Identify the relationships between topics and covariates. For each covariate, estimate its effect on topic prevalence. At the level of individual documents, topic prevalence is the probability that a document belongs to that topic (Roberts et al., 2014). Thus, for each covariate, we are effectively measuring the average topic prevalence of documents associated with each value of the covariate.

Step 5: Interpret the Findings and Reflect on New Insights

Interpret the findings and reflect on new insights that can be gained from them.

Application of the Method

Step 1: Build a Corpus of Survey Responses

We administered a confidential and anonymous online survey to 343 graduate students enrolled in business programmes at an international university on two culturally diverse campuses. Ethics approval was obtained from the institution prior to data collection. A total of 185 students completed the survey, a response rate of 53%. Please refer to Mumtaz et al. (2024) for details on the research project.

The survey included both closed- and open-ended questions. The closed-ended questions included factual/demographic information such as the country of origin and gender. The open-ended questions, among other things excluded for the purposes of this example, asked students to describe how they think ChatGPT and similar AI tools can be used ethically for academic purposes, as well as which country’s culture they most identify with. The responses were collected in English.

In our case, the central question is, “How do you think ChatGPT/other similar tools can be used ethically for academic purposes?”. For the sake of exposition, we focus on two covariates: gender and culture. Other factors may also influence students’ perceptions of the ethical use of AI tools.

The culture attribute ("Which country's culture do you identify with most?") is difficult to organise for the purposes of statistical modelling. Therefore, we drew on the research of Gupta et al. (2002), who classified countries into ten clusters. We used ChatGPT to classify the data on nationality into these ten country clusters by submitting the following prompt to the Large Language Model:

In the following, I want you to classify countries into clusters. These are the clusters:
Anglo Cultures
Latin Europe
Nordic Europe
Germanic Europe
Eastern Europe
Latin America
Sub-Sahara Africa
Arab Cultures
Southern Asia
Confucian Asia

In this prompt, the classification goal is to classify countries into clusters, and the clusters are the possible values (or codes) that the clusters can take. We first tested this with a subset of the responses and made any necessary corrections through follow-up prompts. For example, the following prompt instructs ChatGPT to classify the Philippines as Southern Asia, according to Table 1 in Gupta et al. (2002):

The Philippines should be classified as Southern Asia.

This is an important lesson when using ChatGPT for classification. Its accuracy can be improved by providing additional instructions while interacting with it.

We next asked ChatGPT to classify the replies to the question "Which country's culture do you identify with most?" in the survey data into clusters using the following prompt (this prompt needs to be part of the same conversation as the preceding prompt, which defines how the data should be annotated):

Here are replies to the question "*Which country's culture do you identify with most?*"

Canada
India
Nigeria
Ethiopia
United Arab Emirates
…
Classify these countries into clusters.

Here we prompt ChatGPT with a list of countries from the survey responses and the task to classify these countries into clusters. The results are shown in Table 5.1.

Some responses were undecided (e.g., "Not sure"), could not be mapped to a specific cluster (e.g., "Culture of respect and kindness"), or were blank. For these, ChatGPT assigned the label "No specific cluster identified". Some students also entered more than one country in their response. In those cases (three instances in total), we chose only the first country for inclusion into the structural topic model, assuming that this would be the country the students identified with most.

Table 5.1 Example Output of Classifying Countries into Clusters

Survey response	Code assigned by ChatGPT
Canada	Anglo Cultures
India	Southern Asia
Nigeria	Sub-Sahara Africa
Ethiopia	Sub-Sahara Africa
United Arab Emirates	Arab Cultures
...	...

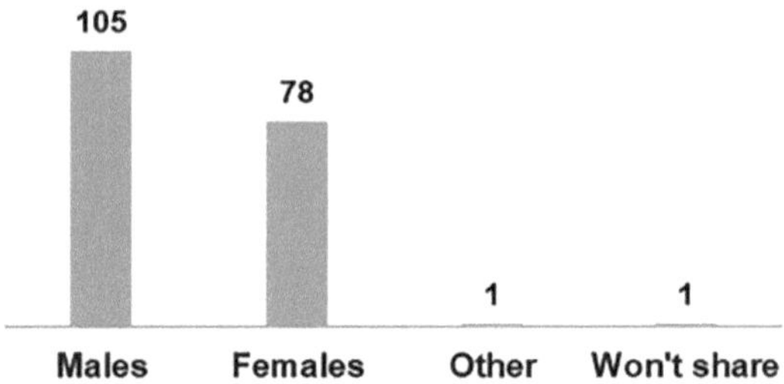

Fig. 5.1 Distribution of Gender

Finally, we manually verified the generated country clusters and corrected any misclassifications. A new column for the computed country clusters was added to the dataset. When using ChatGPT, manual checking is important. Still, the process of annotating the data took less time and led to a more consistent outcome. Using ChatGPT to annotate responses can also scale more easily to larger surveys.

Step 2a: Conduct an Exploratory Data Analysis

For further analysis, the following columns were manually extracted from the survey data: unique ID, gender, country of birth, which country's culture students identify with most, whether they consider the use of ChatGPT ethical, and the new column added for the computed country clusters.

During the exploratory data analysis, we assessed the distribution of each covariate. For gender, the sample consisted of 57% males, 42% females, and 2% other categories (Fig. 5.1).

On culture (country clusters), around half of the sample was derived from Arab cultures, Southern Asia, and Confucian Asia. Nordic Europe had the lowest count among the country clusters (Fig. 5.2). Since there was only one student from Nordic Europe, we considered this as an outlier when we interpreted the results. For all other country clusters, we had a minimum of 7 responses (4%).

Another special case is the cluster of students who did not identify with a specific culture. It consisted of 13 students (7%). Although these students did not provide information about the culture with which they identify, we can get an idea of how homogeneous this cluster is based on the students' countries of origin, which we also collected in the full dataset. When we group the countries of origin into country clusters, we get a distribution that largely mirrors the overall distribution (6 students

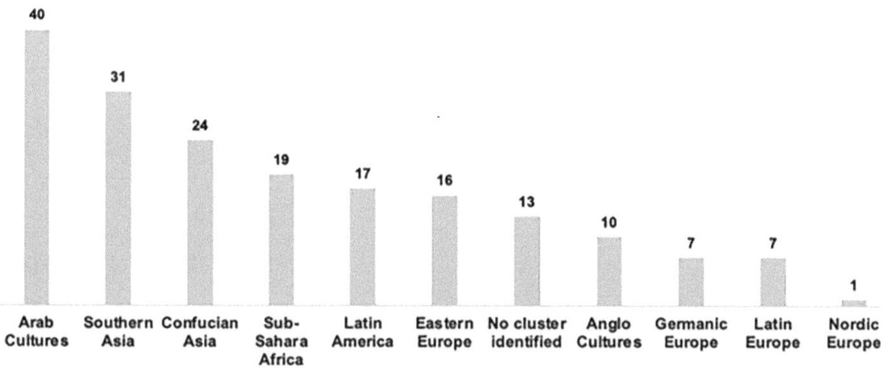

Fig. 5.2 Distribution of Culture (Country Clusters)

from Arab Cultures, and the rest split between Southern Asia, Confucian Asia, Sub-Sahara Africa and Anglo Cultures). Thus, we do not consider this cluster to represent a homogeneous group of students, therefore, we will exclude it from the analysis. Note that we are still showing the cluster in our graphs, as we did not delete the data related to it.

Step 2b: Preprocess the Corpus

Before we could create a topic model from the survey data, the text responses needed to be pre-processed. Pre-processing (converting to lowercase, removing punctuation and common English stopwords, as well as stemming, i.e., reducing words to their root forms) resulted in the removal of 348 terms out of 614, as well as the removal of one document that only contained stopwords. This step is supported by a function of the *stm* package. The final corpus comprised 174 documents and 1417 tokens. These tokens consisted of 266 unique words.

Step 3a: Select the Number of Topics

Given the relatively small number of documents (less than 1000), we expected that a small number of topics would be most suitable. Generally, the more documents the corpus contains, the larger the number of meaningful topics we can have (Weiss, 2021). We generated diagnostic scores within the *stm* package. Figure 5.3 shows the semantic coherence for different numbers of topics between 3 and 6 topics. As we can see from the plot, semantic coherence peaks at three and five topics. Since this metric measures the interpretability of the topics, we initially considered three topics for their higher coherence score. However, we decided to go with five topics because it allowed us to capture a more detailed and nuanced understanding of the data, given the varied cultural and gender backgrounds of the respondents.

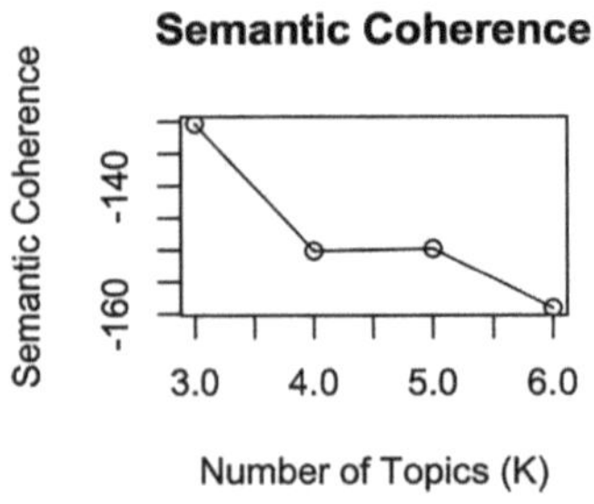

Fig. 5.3 Semantic Coherence by Number of Topics

Step 3b: Label the Topics

Figure 5.4 shows the topic proportions for the five topics. The figure also shows the three most frequent keywords associated with each topic. The proportion of a topic equates to the proportion of documents represented by that topic. Therefore, the topic proportions across all topics add up to 1. According to the figure, topic 3 is the most prevalent, and topic 4 is the least prevalent topic.

From Fig. 5.4, we can see the topics that emerge. For instance, we can infer from the keywords for topic 3 that students perceive an ethical use of ChatGPT as a tool for learning. For a more detailed interpretation, we turn to additional model outputs. Figure 5.5 shows the word clouds for each topic. The size of the words in the cloud is proportional to their frequency in the topic. The word clouds provide a visual representation of the most frequent words in each topic. For example, the word cloud for topic 1 shows that the most frequent words (more precisely, the word stems generated by the preprocessing step) are "use", "question", "inform", "provid", and "subject". In terms of ethical use, these words suggest that ChatGPT is seen as a tool for answering questions.

Table 5.2 shows two of the scoring metrics computed by *stm* for each topic. In our analysis, we focused on the highest-probability words (Highest Prob) and the most frequent but exclusive words, the words that distinguish topics (FREX). The highest-probability words are what we used in Fig. 5.4 to provide initial labels for the topics. The most frequent exclusive words are words that are frequent in a topic but not in the corpus as a whole. For example, for topic 1, the most frequent exclusive words are "question", "subject", "chat". These confirm that the topic is uniquely about using ChatGPT to answer questions about a subject.

By using the word clouds and scoring metrics in combination, we arrived at the topic labels shown in Table 5.3. These represent the students' perceptions of the ethical use of ChatGPT, as revealed by the data. The topic model helped us discover these perceptions.

Step 4: Identify the Relationships between Topics and Covariates

We considered two covariates, as noted in step 1: gender and culture. We wanted to understand if there was a difference in the perceptions of the ethical use of ChatGPT between genders or cultures. Methodologically, it is important to note, however, that

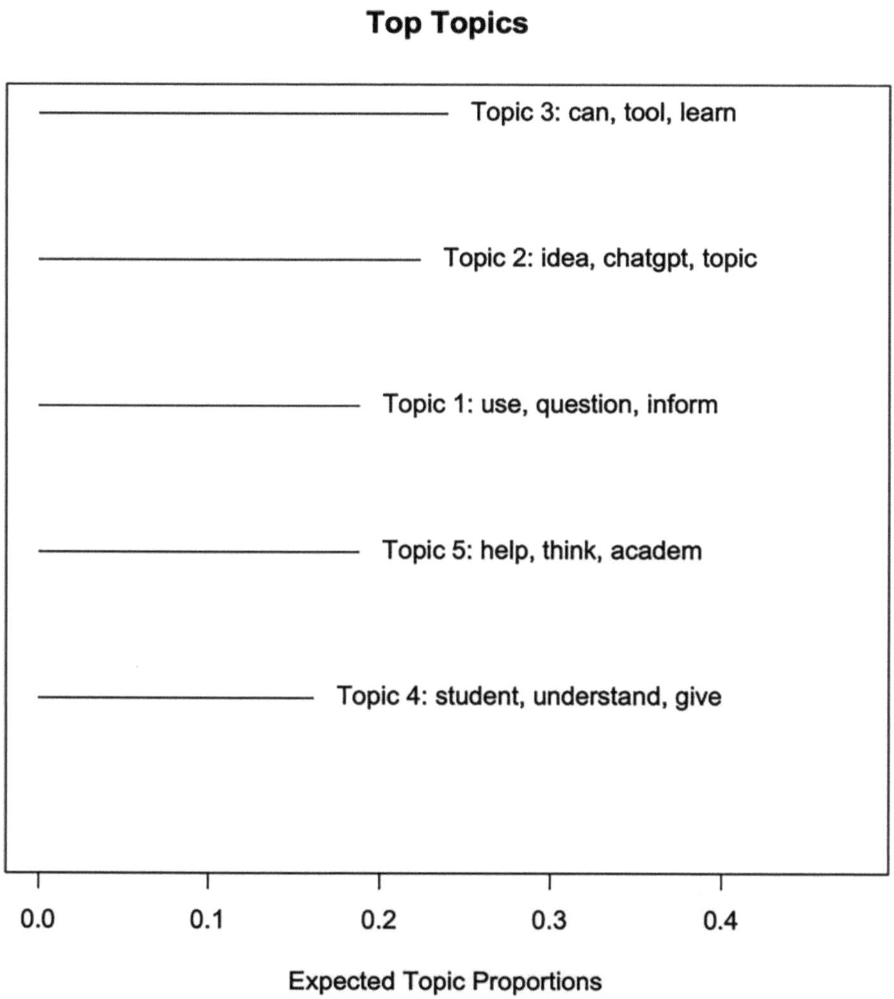

Fig. 5.4 Topic Proportions

the choice of covariates does not impact the topic model itself. What the graphs below show is the topic prevalence for each value of the covariate considered. For example, they can show us whether there is a difference in the topics emphasised by male and female students or not.

Effect of Gender

Figure 5.6 shows the difference in topic proportions between male and female students. The positive x-axis shows topics that are emphasised more by males, and the negative x-axis shows topics that are more emphasized by females. For example, males were more likely to use ChatGPT to answer questions (topic 1) or as a

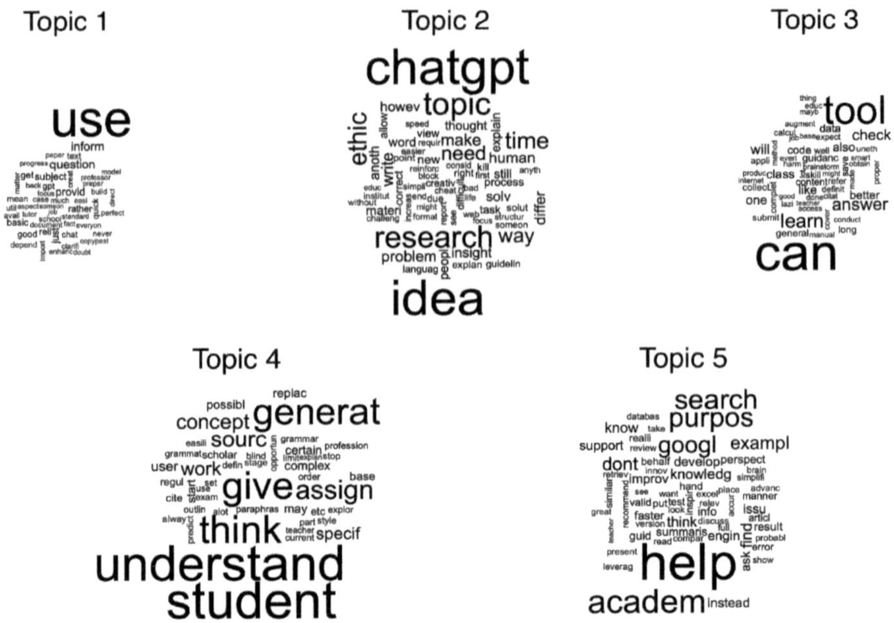

Fig. 5.5 Word Clouds for each Topic

Table 5.2 Topic Scoring Metrics

Topic 1	Highest Prob: use, question, inform, provid, subject, just, get FREX: question, subject, chat, gpt, provid, get, creat
Topic 2	Highest Prob: idea, chatgpt, topic, research, ethic, time, way FREX: make, human, topic, problem, idea, view, write
Topic 3	Highest Prob: can, tool, learn, answer, check, will, one FREX: data, check, like, refer, guidanc, class, long
Topic 4	Highest Prob: student, understand, give, generat, assign, sourc, concept FREX: sourc, work, generat, certain, give, student, complex
Topic 5	Highest Prob: help, think, academ, search, purpos, googl, dont FREX: help, academ, dont, find, know, engin, instead

Table 5.3 Topic Labels

Topic 1	Answer questions
Topic 2	Generate ideas
Topic 3	Use as a learning resource
Topic 4	Understand concepts
Topic 5	Search for information

learning resource (topic 3). Conversely, females were more likely to use it to generate ideas (topic 2), understand concepts (topic 4), and search for information (topic 5).

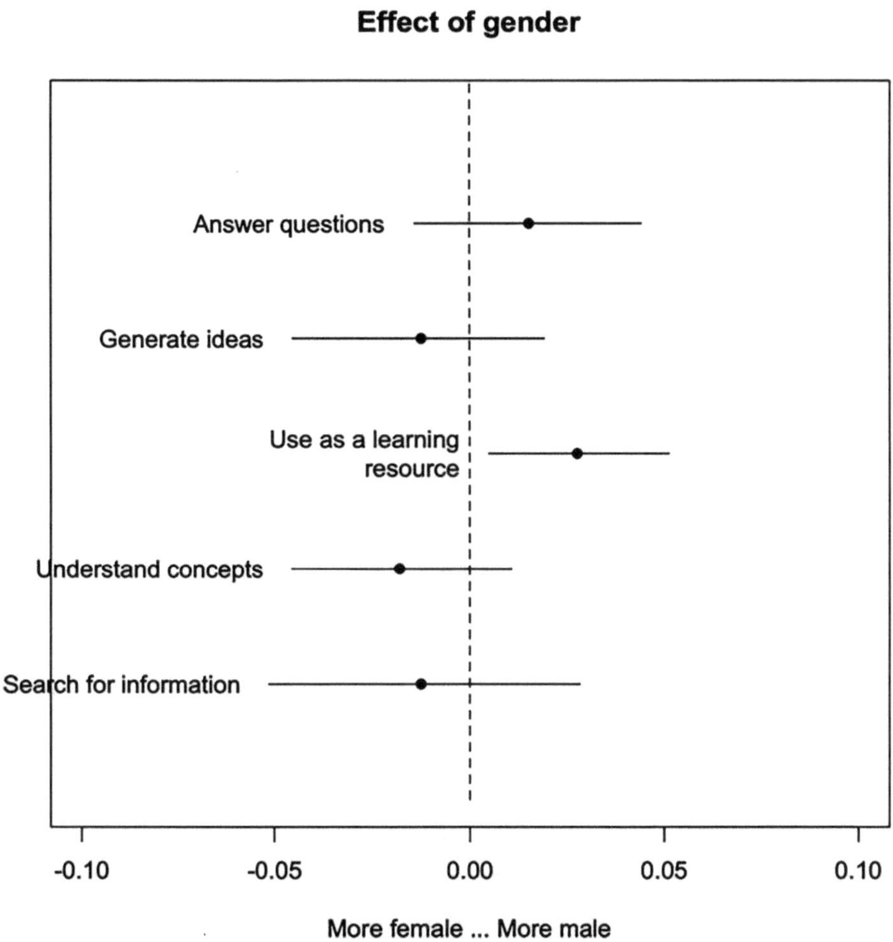

Fig. 5.6 Effect of Gender on Topic Prevalence

Effect of Culture

Analysing the effect of culture on topic prevalence is more complex. We cannot show the effect of each of the ten country clusters on topic prevalence in a single plot. We can only use this approach for binary covariates. Instead, we have presented here the effect of culture for each topic separately and then consolidated the results to understand which cultures have the largest or smallest effect, respectively, for each topic and the variance of the effect across topics.

Figure 5.7 shows the estimated effect of culture on topic 1 (answer questions). The x-axis represents the proportion of students from each culture who mentioned this topic. The figure shows that students from Nordic Europe are most likely to view using ChatGPT to answer questions as ethical, as this cluster has the largest

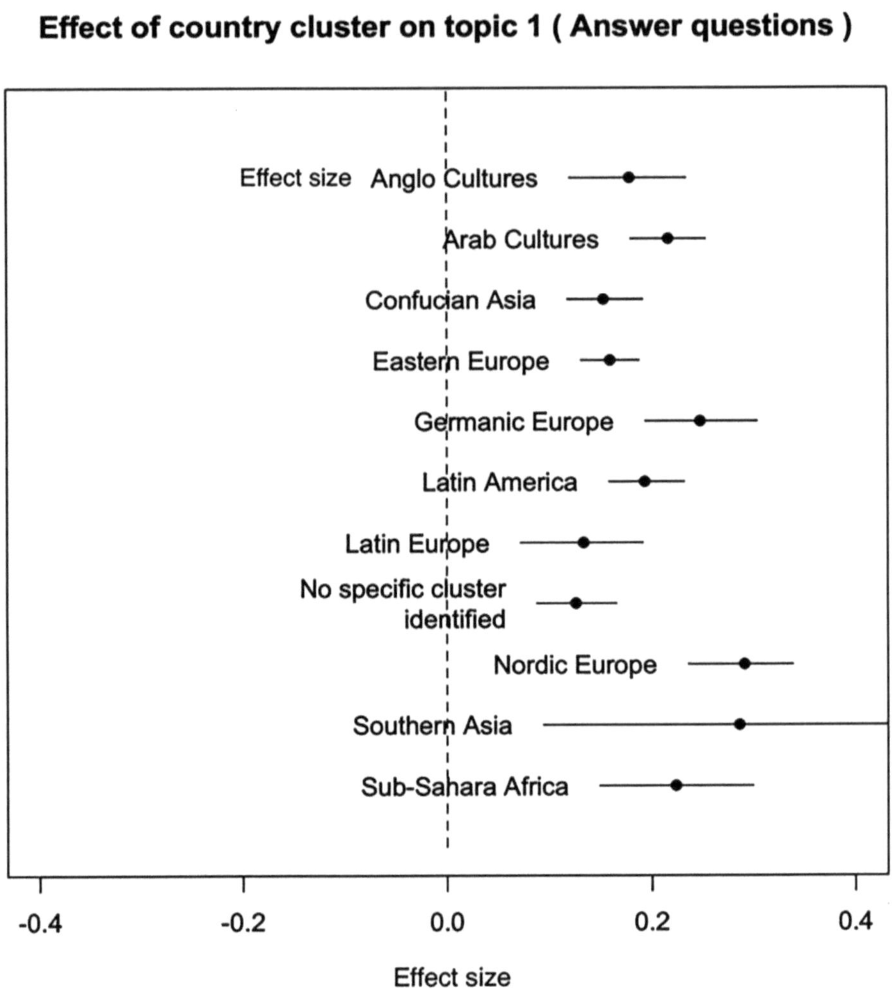

Fig. 5.7 Effect of Culture on Topic 1 (Answer Questions)

effect size. However, since this cluster is underrepresented in the data (as noted in step 2a), Southern Asia ranks next for this topic. In contrast, students from Latin Europe are the least likely to view it as ethical, as reflected by the lowest effect size.

Note that we have also excluded students who do not belong to a particular cluster. As we explained in step 2a, this cluster is not homogenous, and it, therefore, cannot be interpreted as a single culture.

Figure 5.8 shows the estimated effect of culture on topic 2 (generate ideas). Here, students from Anglo Cultures and Latin Europe are most likely to believe that using ChatGPT to generate ideas is ethical (highest positive effect). In contrast, students from Sub-Sahara Africa are least likely to use ChatGPT for this purpose (lowest negative effect).

Figure 5.9 shows the estimated effect of culture on topic 3 (use as a learning resource). According to this plot, the cluster representing Arab Cultures shows the

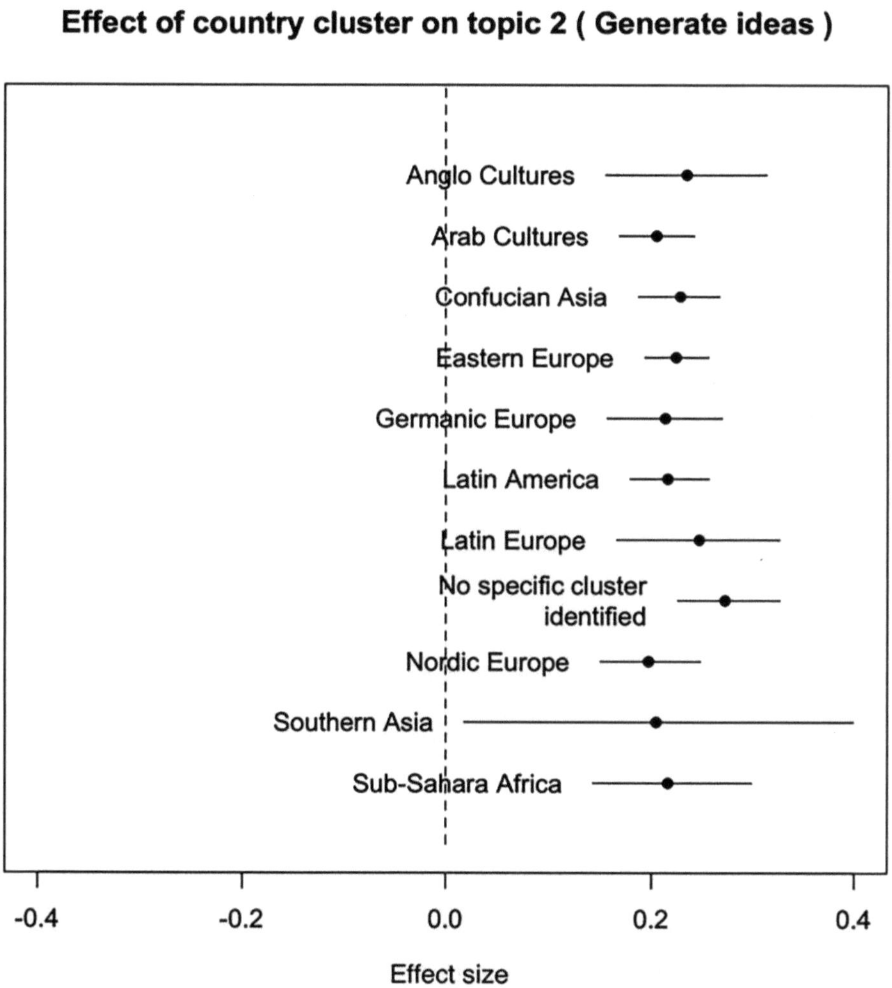

Fig. 5.8 Effect of Culture on Topic 2 (Generate Ideas)

highest likelihood of considering the use of ChatGPT as a learning resource to be ethical. At the other end of the spectrum, respondents from Sub-Sahara Africa are the least likely to view using ChatGPT for this purpose as ethical.

The estimated effect of culture on topic 4 (understand concepts) is shown in Fig. 5.10. For topic 4, respondents from Latin Europe have the largest positive effect size, indicating that they are the most likely to view the use of ChatGPT to understand concepts as ethical. At the other end, the Southern Asia cluster has the smallest effect size, suggesting that individuals from this culture will be less inclined to view the use of ChatGPT to understand concepts as ethical.

Finally, Fig. 5.11 shows the estimated effect of culture on topic 5 (search for information). It can be seen that now, students from Sub-Sahara Africa are most likely to consider using ChatGPT to search for information to be ethical, whereas students from Southern Asia are least likely to do so.

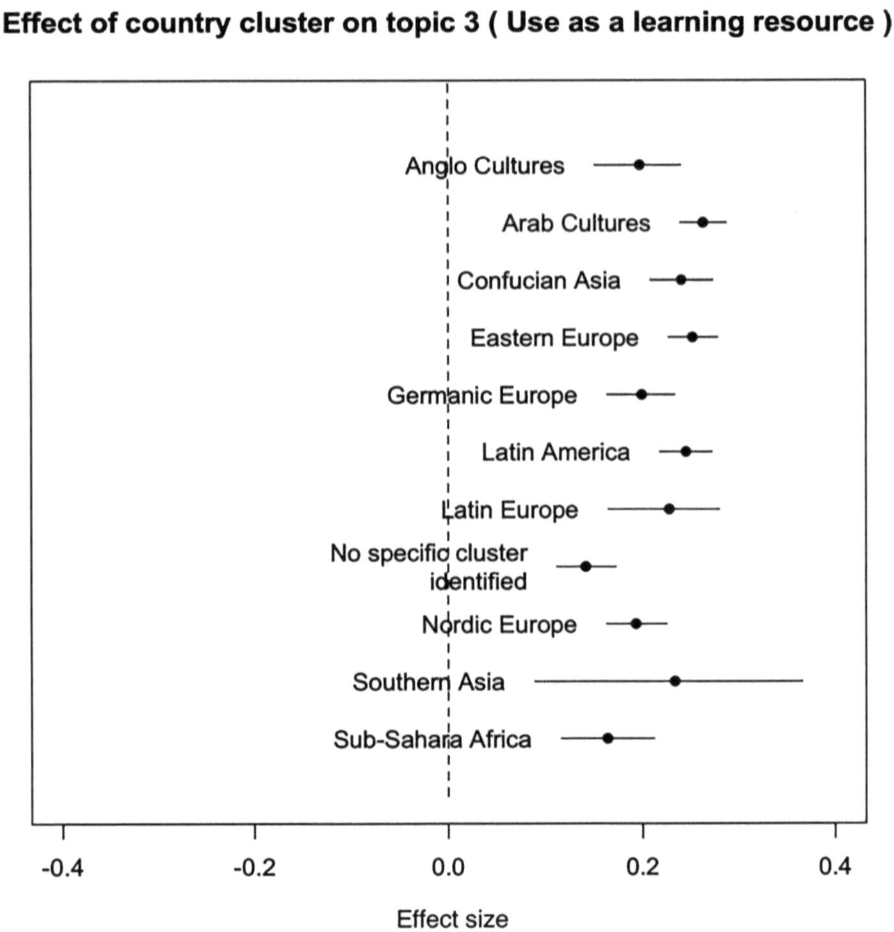

Fig. 5.9 Effect of Culture on Topic 3 (Use as a Learning Resource)

These plots allow us to draw two kinds of conclusions. First, they show which cultures perceive the use of ChatGPT to be more or less ethical for a given topic. Second, they highlight how similar or different the perceptions of ethical use are for each use. A small variance in effect sizes means that views on ethical use are consistent across cultures, while a large variance suggests notable differences between them. Table 5.4 summarises the variance in the effect of culture across the topics.

Table 5.4 shows that topic 1 (answer questions) is the most varied and, thus, the most controversial use case for ChatGPT. Students from different cultures have very different attitudes towards the use of ChatGPT for answering questions. There are notable cultural differences in students' assessment of using ChatGPT as a learning resource (topic 3) and to search for information (topic 5). In contrast, using ChatGPT to generate ideas (topic 2) and to understand concepts (topic 4) is generally seen as ethical across cultures.

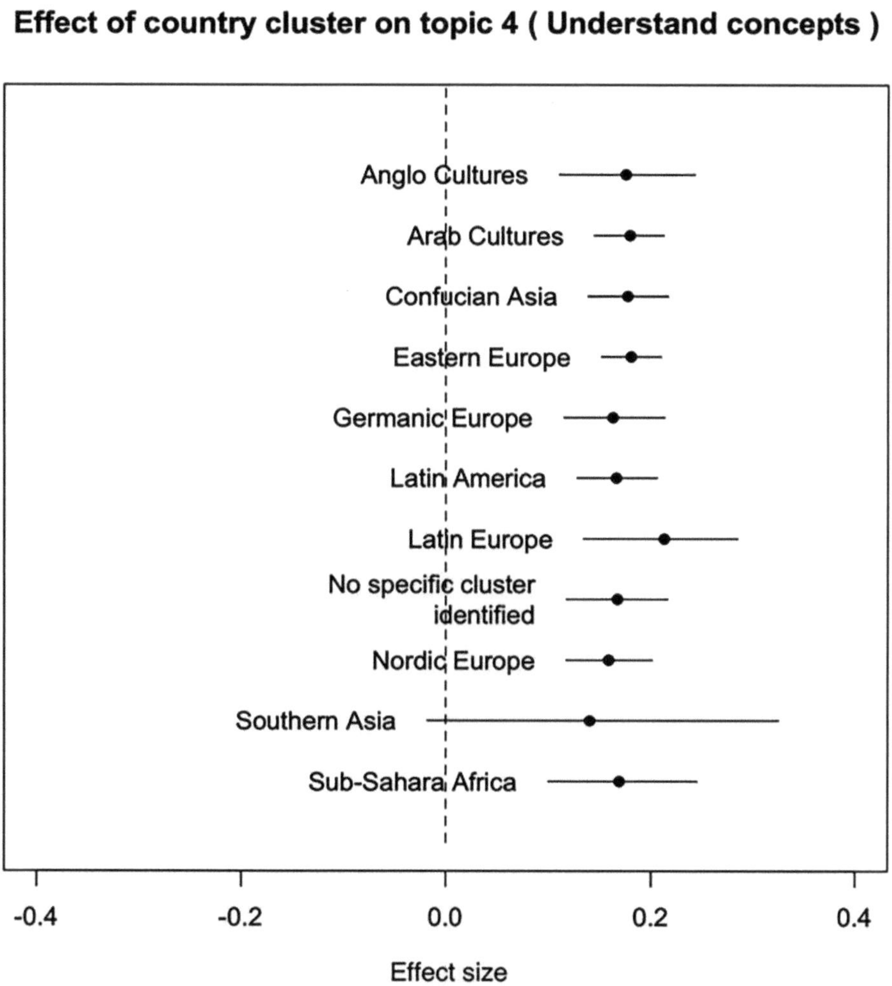

Fig. 5.10 Effect of Culture on Topic 4 (Understand Concepts)

Step 5: Interpret the findings and reflect on new insights

The results of structural topic modelling provide insight into how students perceive the ethical use of ChatGPT for academic purposes. The topics identified in the model represent different uses of ChatGPT. The analysis reveals differences in perceptions across gender and culture. Figure 5.6 shows the differences in perceptions between genders, and Table 5.4 summarises the impact of culture.

We can interpret these differences as reflecting different values between genders (e.g., what is considered hard work or achievement) and the different views on the ethical use of AI tools based on the cultural background (e.g., the importance of the

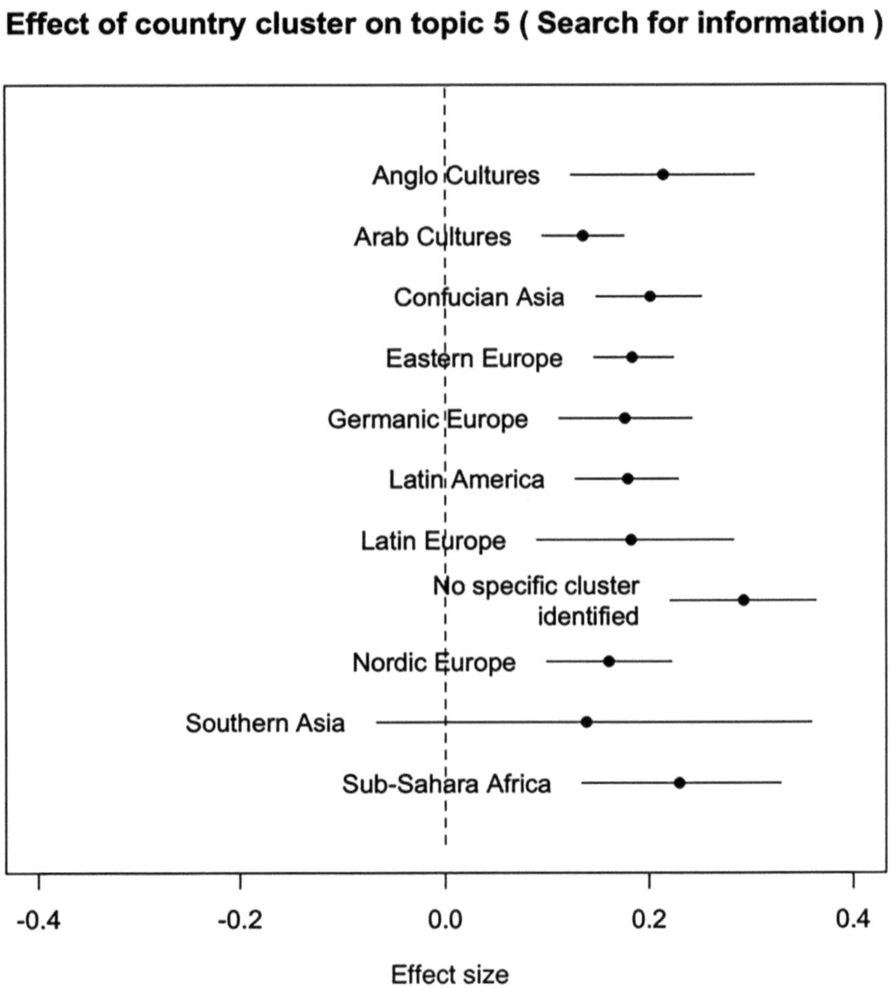

Fig. 5.11 Effect of Culture on Topic 5 (Search for Information)

Table 5.4 Variance of the Effect of Culture Across Topics

Variation in the effect	Topics
Large	Answer questions (1)
Medium	Use as a learning resource (3) Search for information (5)
Small	Generate ideas (2) Understand concepts (4)

individual vs the collective). Please refer to Mumtaz et al. (2024) for a detailed discussion of these differences and their implications.

There may also be other contributing factors that can be analysed by including them as covariates in the same way as the two covariates used to illustrate the approach. These include the level of prior experience with AI and type of learning culture. Their level of familiarity with AI could have an impact on how likely students are to use AI for academic purposes. Similarly, students who have only experienced rote learning may see more value in certain types of AI support than those who have received more interactive forms of learning in their previous education.

Discussion

In line with the primary aim of this study, we focus the discussion on the methodological aspects.

Topic modelling, particularly structural topic modelling, combined with Large Language Models, can help researchers analyse open-ended survey data in a semi-automated manner. When using a topic model, researchers can see the topics or themes in the responses that emerge from the data. The combination of these techniques reduces the need for manual coding and interpretation.

Unlike traditional topic modelling, structural topic modelling also allows researchers to include covariates (such as gender) and measure their effect on topics. Researchers can use this to understand how the perceptions of a topic differ across groups based on the covariates. Another application, not explored here, would be to estimate how topics evolve over time.

As a recent addition to the toolset, Large Language Models (LLMs) can also be used to assist in the analysis of text data. LLMs can help annotate data with no or only a few training examples in a "hybrid" approach to coding with human oversight. This can be particularly useful when working with large datasets or when the data is difficult to categorise.

Of course, no approach is without its limitations. When using topic modelling, some human oversight is still required. In fact, it is desirable. The goal is not to fully automate the process, but to support the researcher in their work. Researchers need to make key decisions throughout the process, from selecting the number of topics to interpreting the topics and assigning labels.

Using LLMs to annotate data is also not "plug and play", but requires careful prompting. The wording of a prompt can affect the outcome, and researchers need to experiment with different wordings until they get the desired result. Human oversight is also important: the output produced by the LLM needs to be vetted.

Conclusion

In this chapter, we described how we analysed survey data on the ethical use of ChatGPT for academic purposes using a combination of structural topic modelling (for identifying themes) and Large Language Models (for annotation). We demonstrated how to incorporate covariates such as gender and culture into the analysis and how to interpret the results. The results provide insights into how students perceive the ethical use of ChatGPT and how these perceptions differ across gender and culture. We hope that this report of our experience will inspire others to take a similar approach to analysing survey data in the area of academic integrity.

References

Blei, D. M., Ng, A. Y., & Jordan, M. I. (2003). Latent dirichlet allocation. *Journal of Machine Learning Research, 3*(Jan), 993–1022. http://jmlr.org/papers/v3/blei03a.html

Carmichael, J. J., & Eaton, S. E. (2023). Security risks, fake degrees, and other fraud: A topic modelling approach. In S. E. Eaton, J. J. Carmichael, & H. Pethrick (Eds.), *Fake degrees and credential fraud in higher education* (pp. 227–250). Springer. https://doi.org/10.1007/978-3-031-21796-8_11

Creswell, J. W., & Plano Clark, V. L. (2017). *Designing and conducting mixed methods research* (3rd ed.).

De Paoli, S. (2023). Performing an inductive thematic analysis of semi-structured interviews with a Large Language Model: An exploration and provocation on the limits of the approach. *Social Science Computer Review, 42*(4), 997–1019. https://doi.org/10.1177/08944393231220483

Gupta, V., Hanges, P. J., & Dorfman, P. (2002). Cultural clusters: Methodology and findings. *Journal of World Business, 37*(1), 11–15. https://doi.org/10.1016/S1090-9516(01)00070-0

Hannigan, T. R., Haans, R. F., Vakili, K., Tchalian, H., Glaser, V. L., Wang, M. S., Kaplan, S., & Jennings, P. D. (2019). Topic modeling in management research: Rendering new theory from textual data. *Academy of Management Annals, 13*(2), 586–632. https://doi.org/10.5465/annals.2017.0099

Heseltine, M., & Clemm von Hohenberg, B. (2024). Large language models as a substitute for human experts in annotating political text. *Research & Politics, 11*(1), 1–10. https://doi.org/10.1177/20531680241236239

Mimno, D., Wallach, H., Talley, E., Leenders, M., & McCallum, A. (2011). Optimizing semantic coherence in topic models. In *Proceedings of the 2011 conference on empirical methods in natural language processing* (pp. 262–272) https://aclanthology.org/D11-1024/

Mumtaz, S., Carmichael, J., Weiss, M., & Nimon-Peters, A. (2024). Ethical use of artificial intelligence based tools in higher education: Are future business leaders ready? *Education and Information Technologies, 30*, 1–27. https://doi.org/10.1007/s10639-024-13099-8

Nelson, L. K. (2020). Computational grounded theory: A methodological framework. *Sociological Methods & Research, 49*(1), 3–42. https://doi.org/10.1177/0049124117729703

Newton, P. M. (2024). Design, run, and interpret survey-based research in the fields of academic integrity and misconduct. In S. E. Eaton (Ed.), *Second handbook of academic integrity* (pp. 1689–1706). Springer. https://doi.org/10.1007/978-3-031-54144-5_182

Roberts, M. E., Stewart, B. M., & Tingley, D. (2019). Stm: An R package for structural topic models. *Journal of Statistical Software, 91*(2), 1–40. https://doi.org/10.18637/jss.v091.i02

Roberts, M. E., Stewart, B. M., Tingley, D., Lucas, C., Leder-Luis, J., Gadarian, S. K., Albertson, B., & Rand, D. G. (2014). Structural topic models for open-ended survey responses. *American Journal of Political Science, 58*(4), 1064–1082. https://doi.org/10.1111/ajps.12103

Schmiedel, T., Müller, O., & Vom Brocke, J. (2019). Topic modeling as a strategy of inquiry in organizational research: A tutorial with an application example on organizational culture. *Organisational Research Methods, 22*(4), 941–968. https://doi.org/10.1177/1094428118773858

Schvaneveldt, R. W., & Cohen, T. A. (2010). Abductive reasoning and similarity: Some computational tools. In D. Ifenthaler, P. Pirnay-Dummer, & N. Seel (Eds.), *Computer-based diagnostics and systematic analysis of knowledge* (pp. 189–211). Springer. https://doi.org/10.1007/978-1-4419-5662-0_11

Weiss, M. (2021). Patterns for the creation and interpretation of topic models. In *Proceedings of the conference on pattern languages of programs* (pp. 1–13). ACM.

Chapter 6
Curriculum Dilemmas: How Can Teachers Foster Pupils' Ability to Collaborate and Compete in the Swedish Education System?

Charlotta Rönn

Abstract The objective of this study is to explore the gap between, on the one hand, the latest Swedish curricula for compulsory school, in terms of the pupils' collaboration, responsibility for their academic success, and their grades, and on the other hand, pupils' everyday life, which is characterised by digital technology and a sharing mentality. This chapter describes curricular features such as self-regulated learning in formal education in Sweden and discusses how these could be understood from the perspective of learner doing schoolwork. The study focuses on how learners, 14–16 years of age, cope with taking responsibility for their own learning in a school setting dominated by constant assessing and grading. Based on classroom observations, audio-visual recordings and interviews with pupils (n = 27), the study examines what learners in the last 2 years of compulsory school at a municipal lower secondary school, do and say when they assist peers with schoolwork without the teachers' awareness. The study shows that the pupils are more loyal to friends than to the formal school system and are willing to forego fair grading as a result. For example, pupils took collective responsibility for individual writing assignments by writing original texts for peers and revising their peers' assignments. Some pupils did not complete much (if any) of the required reflecting and writing themselves. Instead of taking responsibility for their own learning, many saw it as their responsibility to achieve high grades. The study finds that the informal social strategies pupils use when assisting peers reproduce social injustices. The results illustrate the complexity of academic integrity in the juxtaposition between a curricular framework requesting collaboration, individual grading, and the pupils' own responsibility for their learning and the everyday sharing by underage pupils through digital devices. Some consequences are discussed, such as how the current generation of learners, who rely on friends in social networks instead of developing autonomy and self-reflection, might impact future civil society.

C. Rönn (✉)
Linnaeus University, Växjö, Sweden
e-mail: publications@charlottaronn.se

I. Glendinning et al. (eds.), *Ethics and Integrity in Education (Research)*, Ethics and Integrity in Educational Contexts 9,
https://doi.org/10.1007/978-3-031-98157-9_6

Keywords Academic integrity · Curriculum · Digital technology · Grades · Responsibility · Writing assignments

Introduction

In all curricula, there are specifications of what learners are expected to learn as well as ambiguities. However, pupils might also learn what they are not expected to learn, and they can have different attitudes towards learning in line with the hidden curriculum (Broady, 2007; Sundberg, 2021). The latest two Swedish curricula for compulsory school (Swedish National Agency for Education, 2011, 2022) face a major dilemma: pupils are supposed to learn how to write texts both individually and collaboratively with classmates, but they are graded as individuals. The steering documents provide no guiding principle for how the co-written achievements are to be assessed by the teachers. Another dilemma is that school age learners are expected to demonstrate that they take personal responsibility for their own learning. These expectations place the responsibility for actual learning onto minors, who do not have a comprehensive view of the consequences of their actions (cf. Broady, 2007; Carlgren, 2015; Dovemark, 2004; Rönn, 2023a).

It is important to describe and discuss what happens in formal educational settings where learners are to compose texts together with classmates as well as individually, which are graded on an individual basis, and where learners are expected to take personal responsibility for their learning, as is the case in Sweden. In Sweden, as well as internationally, there is an enhanced focus on self-regulated learning (cf. Carlgren, 2015; Hadwin et al., 2017; Zimmerman, 1989), and measurable results achieved by individual learners are common elements in curricula (cf. Rönn, 2023b; Sundberg, 2021). Self-regulated learning has increasingly replaced traditional whole-class teaching in Sweden (Bergqvist, 2012; Carlgren, 2015). The last Swedish curricula for compulsory school (Swedish National Agency for Education, 2011, 2022) emphasise that learners are to take personal responsibility for their studies and simultaneously learn to collaborate with peers and co-write texts with others; the grades, however, are assigned not to the entire group but to individual pupils. Learners are not only expected to take responsibility for their learning, they are also expected to *want to* take responsibility for their learning. The curricula take this one step further by stating that the latter should also be the teachers' point of departure.

In the latest Swedish curricula for compulsory school (Swedish National Agency for Education, 2011, 2022), some of the overall aims are to raise future citizens by preparing pupils for the job market and working life and providing them with guidance as they begin to participate in societal traditions. Internationally, responsibility is a stepping stone in education; the pupils are expected to be active participants in their own learning and take the initiative to acquire skills and knowledge rather than rely on others, such as parents and teachers (Zimmerman, 1989).

According to the Organisation for Economic Co-operation and Development (OECD, 2016), various key competences are regarded as desirable for individuals in educational settings. The concept originates from the business sector, and some of these competencies of individuals are regarded as important from an economic point of view, because they are presumed to enhance market competitiveness, develop a qualified workforce, and contribute to global competition. Knowledge and skills are regarded as important from a social perspective because they are presumed to contribute to individuals' participation in democratic processes, and social justice reinforces autonomy and human rights, as well as decreasing the marginalisation of individuals.

The latest curricula (Swedish National Agency for Education, 2011, 2022) specify that pupils completing the first years of compulsory school are expected to learn how to write texts both together with peers and on their own. Moreover, they are expected to develop the skills needed to provide and receive peer feedback. This is also stated in the comment material on the curriculum (Swedish National Agency for Education, 2017). This focus on both individual and collaborative academic writing skills would be ideally suited for introducing pupils to academic integrity and ethical conduct. However, the curriculum does not specify that pupils learn how to distinguish between creating texts individually and collectively, and teachers are not provided with any specific guidance for how to assess or grade co-written texts in contrast to single-author texts. Moreover, giving and taking peer-feedback on texts is not specifically framed as desirable and appropriate for collaborative writing but problematic when pupils are expected to write independently.

Issues that Arise when Collaborating and Pursuing Higher Individual Grades

The curriculum states that grades should reflect the extent to which the national grading criteria have been fulfilled for each school subject and that the grading process is based on the individual pupil's performance. However, there seems to be an ambiguity that creates a great deal of uncertainty regarding how exactly the teachers are to assess and grade the texts learners produce. Furthermore, because the curriculum does not make a distinction between collective academic writing skills on the one hand and individual academic writing skills on the other hand, learners may fail to make this distinction themselves and use skills that are appropriate in collaborative writing even when they complete individual writing assignments. This may lead to an extensive grey area that can compromise academic integrity. Examples of ambiguous links in the latest Swedish educational reforms are the tensions between competing and cooperating, along with an extensive focus on measurable results and performances (e.g., Erlandson et al., 2020; Erlandson & Karlsson, 2018; Rönn, 2023a, 2024).

Thus, within Swedish formal education, pupils are to work individually and take responsibility for their learning, and simultaneously compete with others in order to get good grades, which are needed for admission to their desired schools and programs. The curricula's enhanced focus and pressure on the individual learner to take responsibility for their own academic success and the grading reforms for compulsory school from 2011 have resulted in increased competition amongst not only learners but also teachers and schools (Löfgren et al., 2017; Sundberg, 2021). The concept of self-regulated learners assumes that learners have both goals and purposes. However, these goals may be opaque and not in line with task goals and objectives that others have set for the learners (Hadwin et al., 2017). It is important to explore and carry out educational research regarding learners' self-regulation of their performances and learning inside and outside the classroom (Rönn, 2023a; Schunk & Greene, 2018). This should be seen in the light of the fact that adolescents regularly share an increasing number of images through social media and value visible feedback, such as "likes". The image-sharing practices of adolescents are complex and related to the development of peer-group norms (cf. Bell, 2019; Winstone et al., 2022) and may offer insights into why pupils engage in cooperative writing practices.

We live in a society characterised by digital technology in which the widespread use of smartphones and social media creates a specific culture of sharing. Today's teenagers have grown up with this; sharing is caring. Simultaneously, the assumption is that learners who take responsibility are able to reflect, work with peers, and possess moral values. In a rapidly changing society where technology is developing steadily and quickly, Ashworth et al. (2003) argue that new moral values and new behaviours (Lindgren, 2015) may appear. Examples of such could be that learners work around the rules they have been given instead of working within them (cf. Rönn, 2023a, 2023b; Rönn & Pettersson, 2023).

It is also important to understand what being encouraged to work with peers but to be graded on one's individual performance might do to the learners' mindset. Having grown up with digital technology, the learners in this age group are used to participating in a culture where sharing is a common practice, which might affect their interest in writing texts individually. One way of approaching this issue is to focus on what the learners do and what they think about collaborating with peers, as well as to focus on their perception of performances in various writing assignments that lead to assessment and grading.

The purpose of the study was to explore and describe the dilemmas within the curricula's collaborative and individualistic aims, how pupils relate to these in their everyday school life, and what the consequences of these dilemmas might be for their preparation for future citizenship.

These are the research questions:

- How do pupils, together with peers, take responsibility for their writing assignments in a school context characterised by a focus on collaboration and individual performances?
- How are the pupils' informal social strategies in assisting peers with writing assignments related to the curriculum?

Method

Gaining Access to the Pupils' Activities Backstage

This study was conducted at a Swedish municipal school, where I followed one class with 27 pupils during their last 2 years of compulsory school in Sweden, i.e. from eighth grade to the end of ninth grade. The pupils were 14–16 years old. To capture the pupils' perspectives, I observed daily classroom activities, documented lessons via discretely staged audio-visual recordings, and conducted semi-structured interviews. Exploring how pupils assisted each other with writing assignments required access to the informal social strategies they applied when they were not under the supervision of the teachers. Informal social strategies were regarded as various casual interactions among pupils; these are not part of the educational rules and expectations, and these may not help pupils achieve the educational aims.

In selecting the targeted school, I scrutinised descriptive demographic statistics for compulsory schools in some Swedish regions with the aim of finding a school where approximately 50% of the pupils had a non-Swedish background (but with few recently arrived in Sweden). Covering more variations in school research enhances generalisability within the targeted research project (Larsson, 2005). At the selected municipal school, about 50% of the pupils were classified as non-Swedish because they and/or their parents were born abroad, either in other parts of Europe, or Asia, Africa or South America. The pupils' parents or guardians spoke 13 different languages apart from Swedish. Guidance notes about the research project were translated into all relevant languages to ensure that all were properly informed about the purpose and implementation of the study. The Swedish Regional Ethical Board reviewed the research plan, comprising observations, audio-visual recordings, interviews and letters of consent (No. 2917/63-31). Ethical approval was also obtained from the school. The pupils, along with their parents and guardians, provided informed written consent for their participation in the study. Thus, all participants were informed. Debriefing took place afterwards.

In the guidance to the pupils and their guardians I explained that the aim of the study was to learn more about how pupils assist each other and that if adults knew more about how pupils assist peers without the teachers' awareness and how pupils think about schoolwork, it might lead to improvements, with teachers adapting their teaching to meet pupils' needs and facilitate the pupils' learning. In order to learn more about their schoolwork, I needed to get access to how they communicated with each other about schoolwork, such as when dealing with formal schoolwork and writing assignments.

The socio-economic circumstances of the pupils varied. Some of the pupils' parents were nurses, teachers and head teachers, while others lived on social allowance. Some pupils lived in villas and had summerhouses abroad, while others lived in blocks of flats. The school's merit rating for Grade 9 was 10% below the national average. Ninety per cent of the teachers at the school were qualified and registered teachers, including the teachers who taught the class targeted for this study. The turnover of the teachers was low; the majority of them had worked at the school for at least 4 or 5 years, and some had worked there for decades.

The core feature of this research project was the communication with classmates and interactions, which normally are not seen or heard by teachers and researchers inside a classroom. The aim was to connect with the pupils and gain their trust so they would want to talk about their schoolwork activities out of the teachers' supervision. The fieldwork was carried out like a three-stage rocket. Firstly, I conducted observations and spent some time with the pupils for several months. I engaged with them during breaks and waited in the corridor with them for the lessons to begin. I had lunch with them in the canteen and observed lessons in all school subjects without participating. I sat at the rear of the classroom, behind the pupils and mainly out of their sight, while observing and taking notes of the pupils' interactions with peers.

Secondly, I placed recording devices inside one classroom and recorded lessons in English as a foreign language, Mathematics, Social studies, and Swedish (mother tongue). The recording equipment was visible but discretely placed around the classroom, focusing on the pupils and not the teachers. For every recorded lesson, the pupils were reminded of the recording equipment before entering the classroom. However, to some extent, the discretely staged recording devices allowed the pupils to become oblivious to them being recorded once the lesson had begun. This is contrary to extensive educational research where the researcher is standing behind a camcorder, visible for everyone to see and a constant reminder of them being observed and recorded. It is common knowledge (cf. Heath et al., 2010) that interactions and patterns of participation may alter to some extent when a person is conscious of being observed and recorded. The fact that I helped pupils forget that they were observed and recorded rendered it possible to get access to the pupils' flow of spontaneous interaction with their peers, such as who talked to whom, what they talked about, and what they did while dealing with formal individual tasks given by the teachers.

Thirdly, the following year, I conducted semi-structured interviews with the same pupils in their last months of compulsory school; four group interviews and 14 individual interviews, comprising 18 individuals, were implemented. Each interview was about 1 h long. The aim of the interviews was to get an insight into the pupils' perspectives on school, on doing schoolwork and why they assist classmates the way they do outside of the teachers' supervision. The interviews consisted of questions such as *How do you proceed when you write an assignment?* and *How do you help each other?* During the fieldwork, I did not ask any questions to the pupils about their grades. I had shown a profound indifference to grades and their measurable performances but focused on connecting with them and trying to understand their perspectives on doing schoolwork. However, during the interviews, the pupils themselves spontaneously spoke about grades, the grading system and how they dealt with having their assignments assessed and/or graded. This led me to ask follow-up questions about issues related to assessed and graded performances.

The results of the study are primarily based on the audio-visual recordings and the interviews, as these changed how I had initially understood the pupils' schoolwork related activities; before, I thought that the pupils worked more independently and that, more often, when interacting with peers, they *explained* exercises and tasks to classmates. In a busy classroom with nearly 30 individuals (comprising the researcher and teachers), many activities occur simultaneously with numerous low-voiced conversations between classmates, which are not supposed to be overheard by the teachers and the researcher. Therefore, I did not capture much of the pupils' informal social strategies by merely observing. The results illustrate *how* peers collaborate when doing schoolwork and the reasons behind their collaborative methods.

The Use of Goffman's Theatre Metaphor as an Analytical Tool for Activities Backstage

How people present themselves to others in everyday life is described by Goffman (1990) through a theatre metaphor. The interaction of a person, together with others, is regarded as an actor who is performing on stage in front of an audience. In engaging in a performance in a specific way, the individual performer intends to give a specific definition of the situation and a particular impression of themselves to the onlookers. Likewise, in a theatre, the intended impression the person wants to give the audience is carried out *frontstage* in the public eye. However, it is *backstage*, beyond the audience's sight, that the person prepares and controls the performances together with peers to prevent the audience from apprehending glimpses of the performance that are not intended to be seen. Only the people on the same team with access to the backstage region are aware of the preparations for performances.

In this study, Goffman's (1990) theatre metaphor is used to describe the pupils' activities outside of the teachers' supervision when dealing with writing assignments given by the teachers, as well as their reflections on school performances and grades. The teacher is regarded as the audience, and the pupils' written assignments that are handed in to the teachers for assessment and grading are regarded as the individual's performance, which the teachers are to assess. The pupils' informal social activities outside of the teachers' supervision take place *backstage* (cf. Rönn, 2023a, 2024).

It is backstage where the learners can prepare a touched-up version of their writing skills. Gaining access to what pupils do and say out of the teachers' and researchers' sight is challenging and time-consuming, both technically and socially. Consequently, research on pupils' activities backstage is scarce (Rönn, 2023a). However, it is of great importance to know more about how pupils prepare their written assignments and how they cope with formal education's demands and pressure upon them. The aim was to get a better understanding of issues related to young learners and to get the background picture of possible issues that might emerge later on in higher education and working life.

A thematic analysis was initially applied, where the transcripts of pupils' actions and conversations were classified as either being intended to be non-transparent (backstage; assisting peers without the teachers' awareness) or transparent (frontstage; aiming at giving a good and intended impression of their knowledge and skills to the teachers). The thematic analysis was connected to the curriculum's aims, such as that pupils should learn how to co-write texts and be graded individually. It was also highlighted what type of future citizen might be fostered in a formal school context characterised by collaboration as well as individual performances that are to be measured and graded. Thus, documentary analysis was carried out, focusing on the content of the latest Swedish curricula and the parts that could be classified as relating to various aspects of the pupils' writing performances and facilitating pupils' (backstage) activities.

Results

Introduction: How the Pupils Assist Peers beyond the Teachers' Supervision

Thanks to the discrete placement of the audio-visual recordings, where the pupils occasionally could forget that their interactions and conversations were being recorded, a new picture emerged of how pupils assisted peers with schoolwork inside the classroom without the teachers' and the researcher's awareness. For example, occasionally, the pupils exchanged computers with peers behind the teachers' backs and wrote original texts for others, logged into their classmates' Google Classroom accounts and wrote original texts for peers and/or made comprehensive revisions of the classmate's texts. In other video-clips, some pupils applied an extensively time-consuming strategy in turning to peers numerous times, instead of to the teacher, when not knowing how to calculate exercises in Mathematics, as well as to systematically copy more industrious and high achieving classmates in chains of copying.

The audio-visual recordings showed how the pupils assisted peers, but the recordings did not reveal why the pupils assisted classmates the way they did. During the semi-structured interviews, the pupils accounted for more backstage strategies and shed light on why the various strategies were applied and which backstage strategies were never carried out at school. Forwarding pictures of completed assignments to peers to be reformulated in their "own words" and sharing leaked National tests on the class Snapchat group were activities that were never carried out at school. As will be illustrated in the following analysis, the grades and the desire to obtain good grades were the hub around which the pupils' ways of assisting classmates revolved.

Switching Computers and Logging in to Peers' Google Classroom Accounts inside the Classroom

One recorded sequence from a lesson in English as a foreign language shows that the pupils are reading a novel in English. After each chapter, there is a set of questions to answer, and the responses to the questions are to be written individually on computers connected to Google Classroom. The camcorder records a sequence where two boys are sitting close to each other, each one with a laptop in front of them. They look at the teacher, and when the teacher turns their back to them, they switch computers with each other. Then, the boy who has a higher level of fluency in English writes the answers to the questions for the other boy. Time passes, and eventually, they look at the teacher again. When the teacher, for a second time turns their back, they switch the computers back.

A few minutes later, the teacher logs in to the Google Classroom account of the boy who has just had the answers written for him, and the teacher gives corrective feedback on the written answers. The boy shouts across the classroom to the teacher that he sees that the teacher has logged into his computer and asks loudly: "Was it good?" The teacher answers that it is good, but that he needs to correct some errors regarding the possessive "s". The boy answers that he will do so. What the teacher in this example is not aware of is that another pupil answered the questions in writing and that it obviously is the other boy who struggles with the possessive "s". Due to one boy's assistance to a classmate out of the teacher's sight, the teacher gives feedback on the text to a boy who has not written that text. Moreover, the teacher reasonably draws a conclusion that the boy who received the feedback had a higher level of knowledge in English than he actually has, apart from the minor errors with the possessive "s". The pupil has, assisted by a friend, managed to give the impression to the teacher of performing better than he would have done on his own.

From the audiovisual recordings, there is a similar example from a lesson in Swedish where three girls are sitting next to each other, working on an individual essay on an author of their choice. They all have a laptop in front of them, and for an observer, it seems as though they are working individually, each one on their computer. However, the recordings reveal that when the girl to the left turns to the girl in the middle (who, of the three, has the most elevated grade in Swedish) to say something, she asks her if she could write a text for her, a text that could be graded with the highest grade, an A. The girl in the middle answers that she is willing to give it a try, but that she first must log out of the computer of the girl sitting on her other side. For an observer in the classroom, such as a teacher or researcher, the impression is that the three girls are working independently on the assignment—each one of them on her own computer. However, the recording devices managed to reveal that they, in various ways, by the use of digital technology, collaborate with the writing assignments in order to get (more) elevated grades. Later on, the teacher reads and gives constructive feedback to the girl who has just had her friend attempting to write an A-text for her. The teacher cannot know that it is not this girl who has written the text. These examples show that the impact of the formative assessment is sidestepped; the teachers give formative assessment to pupils who have not written the texts, instead of addressing the feedback to pupils who did the work (but without the teachers' awareness). The teacher gives appropriate formative assessment to pupils who have written *their own* texts, but the feedback to pupils who have had all or part of their assignments provided by peers will not correspond to their learning needs. Not only do they not exercise reading, reflecting and writing—their feedback does not reflect their present stage of development, or guide them on how to proceed and/or improve their skills.

The audiovisual recordings, moreover, showed two different examples of when several pupils are involved in carrying out a task given by the teacher. One example is from a lesson in Geography. The recordings showed a sequence from a lesson where all pupils were asked to read a few pages in the textbook and then answer some questions on irrigation in Palestine and Israel. One girl with good grades in Geography started reading the text and answering some of the questions in her copybook. Once she had answered a set of questions, she passed her copybook to the girl sitting on her right while she went on reading more and preparing for more questions to answer. The girl next to her copied her answers, then handed her own copybook (with the copied answers) to the girl on *her* right and handed the copybook back to the first girl. The entire procedure was repeated twice during the lesson, and while all three girls ended up with answers to questions that they had written in their own copybooks through chains of copying, it was only the first girl who actually had bothered to read the text, and according to her understanding of the text, answered the questions. As an observer in a classroom, it is possible to see occasional interactions between classmates where they seem to copy someone's answers. However, it required the discretely staged recording devices to map out how systematically and casually the copying can occur.

One example of time-consuming assistance among peers, instead of turning to the teacher for help, is from a lesson in Mathematics, where one girl tries to solve an exercise to which she does not have the answer key. The task is calculating percentages, and she has calculated the answer as "one". However, she does not think her answer is correct, but instead of seeking assistance from the teacher, she seeks assistance from the girl sitting next to her. The girl beside her calculates the exercise twice in different ways and gets the result "one" both times, but she does not think it is correct either. Instead of turning to the teacher, the second girl proposes to turn to the girl in front of them. This girl, too, calculates the exercise and gets the result "one". Now that they have started to discuss whether "one" could be equivalent to 100%, they agree that this is the case and that the calculations they have made are all correct. It is not until now when the girls know the correct answer that the first girl turns to the teacher. Instead of asking how to calculate the exercise, she now tells him how to do it, and she also explains that "one" is equivalent to 100%. The girls do not turn to the teacher when they do not understand. Instead, they turn to the teacher when they do understand. A few minutes later, another girl turns to the three girls who have just been involved in solving the exercise; she does not seek assistance from the teacher, either, but from friends who could explain how to proceed in calculating this kind of question. Towards the end of the lesson, the first girl had calculated or started to calculate the same exercise ten times. It seems to be an example where integrity is not compromised, even though it is a very time-consuming strategy compared to turning to the teacher and asking for help. In the interviews, it became clear that the pupils often preferred to keep the teacher from knowing when they did not understand.

Some of the informal social strategies the pupils applied together with classmates in aiming for good grades with little effort were a) switching computers with a friend, b) logging into friends' Google Classroom accounts and writing texts for peers, c) applying chains of copying someone's assignment and d) applying chains of pupils who explain to peers. Out of these strategies, only the latter tends to lead to autonomy and, in a long perspective, taking responsibility for one's *learning* and not solely for having a written assignment to hand in for grading. In doing so, it also matches the intentions of the curricula, while the others might be regarded as unintentional side effects of the curriculum, which aim to encourage pupils to collaborate on texts together with peers.

Forwarding Pictures of Completed Assignments and Sharing Leaked National Tests outside the Classroom

In the interviews, the pupils accounted for more strategies, and it became apparent that the informal assistance inside the classroom was merely the tip of an iceberg. The pupils explained that in evenings and weekends, some pupils turned to more industrious and/or high achieving peers and asked them to send screenshots of their completed writing assignments. Once the classmate had forwarded an image, the requesting pupil rewrote the text in their "own words", for example, with the help of a website specialising in paraphrasing.

In an individual interview, one girl explained that she could receive up to three requests from different classmates regarding forwarding the same assignment and in a group interview with several high-achieving girls, they said that once one had decided to forward a picture of an assignment to someone, this person could thereafter forward it to others so the one who had written the original text lost control over who was reformulating it. The pupils took shared responsibility when peers had a writing assignment to hand in for individual assessment or grading—an assignment that several pupils had worked on to various degrees.

During 16 out of 18 conducted interviews, the pupils spoke abundantly about the extensively leaked National tests and the leaked teachers' grading criteria. It became evident that the printed tests on paper (which obviously had not been kept satisfactorily secure by adults at schools around the country) had been leaked more extensively than the Swedish National Agency for Education had informed teachers and the public about through their website and mainstream media. With the leaked tests and leaked teachers' grading criteria, the pupils could practice beforehand on the greater part of the 15 tests held in various school subjects during the last year of compulsory school. In the interviews, the pupils explained that they could memorise multiple choice answers, practice beforehand in writing essays once they knew the topic and learn the correct content for answers in freehand writing. Moreover, with help from the teachers' leaked assessment instructions, the pupils could choose in advance to opt for a grade that roughly corresponded to their current grade, which reasonably prevented the teachers from discovering the extensive leaks.

The pupils disclosed that the leaked tests spread nationwide, but no one ever admitted that it had spread from their own school. The pupils explained that the first pupil in the class who got access to a leaked test and/or leaked teachers' grading criteria shared it with classmates in the class Snapchat group, where all the pupils had access, although some had chosen not to partake in the sharing of the leaked National tests. In the interviews, the pupils explained that they regarded adults as the ones responsible for the leaked National tests because it was adults who had not safeguarded them sufficiently before the test day and thus adults who leaked them. Adults' backstage activities, such as leaking the National tests, rendered it possible for some pupils to share the leaked materials with classmates backstage and to improve their individual results on the National tests.

The Pupils' Regard on Schoolwork

The pupils applied various informal strategies where they cooperated with peers without the teachers' awareness in backstage spaces; they:

- switched computers with peers who wrote original texts for them;
- logged into peers' Google Classroom accounts inside as well as outside school and more high achieving pupils wrote original texts and/or carried out considerable revisions for peers;
- copied peers in chains of copying;
- explained to peers in chains of explanation instead of turning to the teachers for assistance when they did not understand;
- forwarded images of more high achieving classmates' completed writing assignments for others to reformulate in their "own words"; and they
- shared the extensively leaked National tests in the class Snapchat group.

In conversations with the teachers after the pupils had left compulsory school, the teachers said they were dismayed at the pupils' social strategies in doing schoolwork. To some minor extent, they suspected that it occasionally happened that pupils somehow worked on writing assignments out of their supervision, but the various ways of doing this, the scope of it, and how digital technology facilitated the sharing of assignments were new to them. The school used Urkund (later Original, which is now part of Turnitin) for plagiarism control, and the pupils' writing assignments passed with few exceptions. The pupils claimed concordantly that the teachers had no idea about their extensive collaboration.

By engaging in the previously mentioned activities where the pupils assisted classmates without the teachers' awareness, the pupils sidestepped two control mechanisms for just and equitable grading; first, on a local level, where the teachers were kept from knowing about how some of the pupils assisted classmates in ordinary schoolwork, and also on a national level where some pupils profited from leaked and shared National tests. It can be argued that if all pupils had equal access to peers' assistance nationwide, there would be a sense of equity in grading. However, this was not the case; inequalities emerged that were related to language mastery in the language of instruction (Swedish), socio-economic features, and also gender. More than ten different first languages were spoken by pupils in the class, and those pupils who were less fluent in Swedish did not seem to have equal access to assistance from the high achieving pupils. For example, it would be difficult for someone fluent in Swedish to write a convincing original text for someone who struggles with the language. Due to socio-economic factors, some pupils did not have their own computers or access to the internet at home, which prevented them from partaking in having texts written or forwarded. In addition, some pupils did not have telephones that could receive images, which excluded them from receiving pictures of completed assignments from the high achieving pupils to be reformulated digitally outside school. High achieving girls explained that it was easier to turn down requests from boys than from girls; if requests from girls were rejected, the rejected girls responded by socially punishing the high achieving girls by ignoring them at school for a month or two, but if a request from a boy was turned down, no punishment followed. As a consequence, the girls more often turn down requests from boys than from girls. The teachers' grading criteria for the National test are designed for teachers. Consequently, one needs to have good cognitive and language mastery in order to benefit from more in-depth instructions regarding content and assessment. In the interviews, pupils who struggled at school explained that they did not manage to make much sense of the leaked teachers' instructions. Classic social patterns related to language mastery and socio-economic features were reproduced and reinforced by the pupils' interaction with peers backstage, which generated inequalities between the pupils.

These inequalities were related to the individual pupil's access to high achieving informal networks and classmates willing to share texts and write original texts for them. This was facilitated on a large scale through the computers and the software, Google Classroom, which were provided by the school, as well as the pupils' everyday use of smartphones. Sharing written assignments was prevalent. Moreover, pupils with an extensive social network had better opportunities to improve the written assignments handed in to the teachers for individual assessment or grading. In interviews, the pupils explained that they had started sharing images of completed written assignments in middle school, the first year they were graded, but at that time, they did not understand the consequences of bringing their everyday sharing culture into formal schooling. The pupils gave a concordant picture of schoolwork being difficult and that the exercises constantly got more and more difficult. The pupils had started to get worried. In the interviews, several pupils said that when they started sharing completed pieces of homework in sixth grade, they were not aware that they were not meant to do this. It wasn't until eighth grade that they fully understood that this was not how schoolwork was meant to be carried out, and some pupils tried to stop sharing their writing assignments with classmates. However, no one amongst those who were assisted by peers described having started to do their own writing assignments.

According to the pupils, relying on peers who were willing to assist became a vicious circle which proved difficult to stop. In an interview, one girl who had been assisted by peers during the last 4 years of compulsory school said she was worried about what would happen after the summer when she started upper secondary school, because the teachers there would believe that she had knowledge and skills matching her grades—but she did not, and she would not have her old classmates to rely on either. One girl, who previously had provided assistance with writing assignments to classmates but who had ceased to do so in previous years, said ironically, "Good luck in the future!" when talking about pupils who relied on peers' assistance instead of doing their assignments themselves. She meant that depending on others was a poor strategy for higher education and working life.

In the semi-structured interviews, it became obvious that the pupils wanted to make a good impression on their teachers and felt it was important to perform well. Additionally, peers assisted each other in order to have good writing assignments to hand in and get good grades. *The grades* turned out to be both the triggering point to put the pupils' collaborative activities in motion and simultaneously the results of the students' assistance backstage.

The Curriculum and How the Pupils' Activities Relate to its Aims

The overall aims of the latest Swedish curricula (Swedish National Agency for Education, 2011, 2022) are to contribute to the pupils' individual development based on their individual prerequisites, foster future citizens in democratic values and societal traditions, and prepare them for the future job market and working life in a rapidly changing global world. The curricula further state that the pupils must take personal responsibility for their learning and be willing to take responsibility for their learning. The teachers should start with the understanding that the pupils want to take responsibility for their learning. Not only is the individual pupil required to take responsibility for their learning, but the pupil should also learn how to work with classmates. In the latest Swedish curricula (Swedish National Agency for Education, 2011, 2022), it is stated that pupils are to collaborate and learn how to write texts together with others, as well as to give and take feedback on their own as well as their peers' texts. The results of this study show that pupils seek assistance from their peers in producing texts to hand in to teachers as individual accomplishments within a school context that heavily relies on formative and summative assessment. The difference between formal education now and formal education before digitalisation is that today, social media, software such as Google classroom, and the pupils' everyday use of smartphones to share pictures of written assignments, make it possible for pupils to share work and assist peers on a much larger scale and with more ease, even to the point that some pupils write original texts for their peers.

All the activities the pupils carry out without the teachers' awareness relate to different parts of the curriculum. Embedded in the curricula are social strategies that pupils apply in their backstage pedagogy in responding to formal education's demands, such as producing and providing texts to be assessed. Some examples of what the curricula state compared to how the pupils respond to this in backstage activities are:

- From their first years in compulsory school, pupils are required to co-revise and co-create texts with classmates. Surprisingly, there are no instructions to the teachers of *if* (and in such cases, *how*) these texts are meant to be graded individually, which creates a grey area for assessing and cooperation. The pupils, with their sharing mentality and use of digital technology since infancy, collaborate with peers, write original texts for one another, revise each other's texts without the teachers' awareness, and reformulate their peers' completed writing assignments.
- The digitalisation of the educational system requires schools to invest in computers and software, such as Google Classroom, in order to prepare pupils for future societal demands. However, today's pupils, who have grown up with smartphones and digital technology, bring a *sharing culture* to school work and apply digital technology, such as social media and informal networking, at school in the same way as in their everyday lives.
- An enhanced focus on formative assessment in terms of scrutinising where the pupil is at, where (s)he is heading, and how (s)he is going to get there. According to the curricula, the head teachers and teachers are obliged to meet the pupils' current educational needs. However, by logging into classmates' Google Classroom accounts and writing texts for them, outside of teachers' supervision, the pupils actually hinder head teachers and teachers from mapping out the pupils' actual educational needs.
- Numerous National tests are part of summative assessment to ensure the principle of equitable grading of each pupil's level of knowledge. However, the pupils collaborate on these texts, too, by sharing tests leaked by adults in the class Snapchat group and nationally across schools. This allows the pupils to improve their results without the teachers' awareness, which undermines equitable grading.
- That formal education aims at equality between pupils regardless of the pupils' prerequisites and skills. However, the pupils' activities backstage reproduce injustices related to language mastery, socio-economic features and gender.
- That the pupils must take responsibility for their learning and should be willing to do so. However, as it turns out, many pupils are rather willing to take responsibility for having an assignment to hand in to the teachers for grading, thereby

giving the impression of performing well—rather than engaging in the long and tiresome process of learning and acquiring skills and knowledge.

The findings show that all the pupils' activities and assistance backstage mirror, in one way or another, what is imposed upon them through the curricula. A complex and troublesome issue emerges when a generation of pupils who have grown up with tablets, smartphones and social media, with a sharing culture as a natural part of their everyday life, meet the formal educational context, where: a) grades are important, b) pupils should learn how to work together on texts without pertinent instructions to the teachers if, and in such case how, these texts are to be assessed, c) pupils should be responsible for their academic success, and d) digital technology is provided by the schools. This seems to result in what might be considered misconduct based on misunderstandings and unclear frameworks, which affects the quality of formative feedback received by some pupils and, as a consequence, their learning.

In this blurred educational and societal context, the main question is not whether the pupils' behaviours are ethically or morally acceptable. It is important to determine how adults (e.g. policy makers, teachers, parents, researchers) can support each pupil's learning, and how they can prevent young learners from sharing assignments in a way that leads to some pupils missing out on learning opportunities. Concerning the immediate future of this current generation of pupils, it is relevant to consider what the possible consequences of the mismatch between curricular aims, the digitalisation that facilitates for learners to large-scaled assistance of numerous peers, along with the widespread sharing culture, could be for higher education and working life. Maintaining academic integrity in higher education could be difficult in such an everyday culture of using smartphones, social media and sharing, especially where some students have had similar experiences to those of the pupils in this study.

Discussion: What About the Development of Future Citizens?

Today's pupils are expected to take personal responsibility for their learning in line with self-regulation. However, these pupils are not in a position to fully appreciate the long-term consequences of their actions when providing or receiving peer assistance during their learning and assessment activities. Since their early childhood, young people today have been accustomed to managing smartphones and tablets and sharing pictures of their everyday lives with family, friends and acquaintances. Naturally, some extend these behaviours to their schoolwork. However, the *learning process* for some pupils becomes ineffective and lacks integrity when schoolwork is outsourced to more high achieving and industrious peers who are willing to assist.

In this study, some of the pupils were loyal to their classmates, revised peers' texts, wrote (without being paid) original texts for others, forwarded images of their own completed writing assignments to be reformulated by their peers, and shared leaked National tests on the class Snapchat group without the teachers' awareness. The pupils' use of digital technology and their reliance on informal networks facilitated these activities, also on a large-scale, in backstage spaces, which resulted in a product to hand in to the teachers rather than developing the individual learner's academic process and integrity.

Internationally, likewise in Sweden, formal education is influenced by new public management ideals that emphasise transparency and effectiveness. Learners are required to develop competences that are presumed to enhance productivity and competitiveness, which are regarded as important features in the global job market in the future. This study showed that an emphasis on transparency and effectiveness (such as the assessing, grading and measurement of individual performances) can lead pupils to make their schoolwork activities less transparent for the teachers by moving them beyond the teachers' supervision.

With the results of this study as a starting point, it is important to reflect upon a) the impact of the current policy documents in many countries around the world that underline transparency and effectiveness, b) the personal responsibility for their learning imputed to the children in an educational context where accuracy of measurable results are considered paramount, c) that pupils should learn how to write texts with peers, but without clear instructions to the teachers regarding assessment, and d) the sharing culture children have developed since their early childhood by their use of smartphones, social media, and formal education's use of software for sharing documents in progress. It might become problematic on several levels when individuals who already in compulsory school learned to rely on shared assignments without integrity, pass through the educational system, and enter the job market. A very different type of future civil society may emerge when the current generation of learners comes of age; they who rely on informal strategies and networks and who recycle others' content and arguments instead of expressing their own opinions and making their voices heard. The results of this study may help explain the origin of some issues related to academic integrity, misconduct and cheating behaviour that occur in higher education and emphasizes the importance of addressing the issues of merit based achievements and (large-scale) providing and sharing of original texts for others already in compulsory school.

It is important to reflect on how the curricula have been implemented and how they have enabled some of the future citizens to become dependent on informal networking and activities conducted backstage, while simultaneously giving the impression of taking personal responsibility for their formal work tasks in line with the current presumed individual skills and knowledge. Although not observed in this study, in our rapidly changing world, it is not solely peers who can assist pupils backstage outside of the teachers' supervision. Parents, siblings, and friends outside school, as well as artificial intelligence tools such as ChatGPT and other chatbots, can similarly write original texts and/or revise learners' writing assignments without teachers' awareness. Thus, the various backstage options for informal assistance without the teachers' (or future employers') awareness make it essential to question what flexibility and competence this future workforce will possess if large-scaled backstage activities facilitated by digital technology go unnoticed. It is important to question what impact this might have on the development of young people's awareness of democratic processes, social justice and equity principles, which are highlighted in the aims and objectives of the curricula.

It seems as though encouraging cooperation between peers in composing texts and simultaneously highlighting the importance of measurable performances and grading of the individual's writing assignments *without giving clear curricular instructions* of how this is meant to be carried out has facilitated the emergence of comprehensive assistance between peers backstage, lacking transparency. In addition, there is a widespread sharing culture in society, which educational institutions need to handle and adapt to. More research on how learners relate to, and deal with, formal educational demands backstage is urgent, likewise to ensure that the individual pupils develop skills and abilities to express *their own* opinions as well as engage in the time consuming task of *learning*—not merely quick fixes such as collaborating with peers backstage in producing assignments for assessment.

It is important to bear in mind that some activities carried out outside of the teachers' supervision are constructive and promote learning, problem solving and autonomy which the example of calculating exercises in Mathematics' shows. However, it seems that the majority of activities backstage are not constructive from a learning perspective, as they hinder the learning processes for some pupils and keep them from developing problem-solving skills and autonomy in line with the curricular aims.

The findings of this study show a complex reality defined by a) formal educational norms where grades are regarded as important, b) an everyday sharing culture amongst learners, which is facilitated by digital technology, and c) minors who must take responsibility for their learning even though they cannot foresee the consequences of their actions. The clash between these features, full of inconsistencies, opens up a breeding ground for questionable values and behaviours, which may have consequences when these young people enter higher education. One thing learned from this study is that in today's digitalised society, learners bring with them a sharing culture into formal education, which has consequences for the pupils' learning and the teachers' assessments and grading. It seems questionable whether the pupils' activities outside of the teachers' awareness align with the goals set by the latest Swedish curricula (2011, 2022) and the OECD (cf. OECD, 2016), such as developing a qualified workforce and contributing to social justice and democracy, while also meeting another goal of the OECD: enhancing market competitiveness.

The purpose of this study has been to show the complexity of the gap between the latest curricula's emphasis on collaboration and individualistic aims and today's pupils' everyday life while doing schoolwork, as well as the consequences this might have for preparing pupils for future citizenship. A dichotomy between, for example, moral and immoral behaviour in compulsory school seems overly simplistic, seen in light of the complexities surrounding collaboration, individual grading, pupils' personal responsibility for their learning and the sharing culture of the younger generations, which is facilitated by their everyday use of smartphones and social media. What is useful is to regard this not as an issue related to the individual child but as a societal reality and seek to find perspectives to promote time consuming learning processes amongst *all* learners (and not just those who already tend to do this and who are the ones who share their completed assignments with peers). More research is needed, but a hypothesis is that the learners' activities highlighted in this chapter might lead to a future society where many citizens might lack independent problem-solving skills. This study shows the complexity of academic integrity concerning young learners who are legally required to attend school, some of whom are avoiding learning opportunities and not taking responsibility for their own learning. It goes without saying that this is in sharp contrast to adult students who have chosen to apply to higher education and who are able to foresee the consequences of their actions.

For some of the pupils, taking responsibility for their learning bears the stamp of social media and large-scaled sharing. One important issue to discuss is whether it would be possible to minimise the non-constructive activities outside of the teachers' awareness without affecting the Swedish National Agency's guidelines for collaborative learning and the pupils' expected willingness to take personal responsibility for their learning in everyday life. As a result of this study, at the school where the study was conducted, teachers now insist that pupils do their writing inside the classroom (even though they might prepare for the writing at home) to provide more certainty about who has written what. When the writing is done in class, teachers can address the needs of individual pupils and offer adequate assistance tailored to each pupil's needs. Moreover, the teachers now rely more on oral assignments and handwritten assignments since the results have been reported back to them.

More research is needed, but I will end this chapter by making a few suggestions that might contribute to minimising the learners' non-constructive backstage activities:

- Talk to learners about sharing cultures in educational settings (advantages and disadvantages) seen from both short-term and longer-term perspectives. Discuss with pupils whether sharing is due to caring, peer pressure and/or some degree of social coercion.
- Adjust the assignments to the pupils' different development stages and design individualised assignments to prevent large-scaled sharing.
- Provide a clear local framework for co-written texts and whether or not these assignments are to be assessed.
- Limit the use of platforms like Google Classroom to working on collaborative assignments; reassess whether all documents and assignments should routinely be shared on such platforms.

Last but not least, engage learners in discussions about the difference between *information* and *knowledge*. Information is just a click away and can easily be copied and/or shared, such as completed writing assignments. Knowledge takes time and effort to achieve, such as knowing how to write one's own assignments and to express one's own opinions.

References

Ashworth, P., Freewood, M., & Macdonald, R. (2003). The student lifeworld and the meanings of plagiarism. *Journal of Phenomenological Psychology, 34*(2), 257–278. https://doi.org/10.1163/156916203322847164

Bell, B. T. (2019). "You take fifty photos, delete forty nine and use one": A qualitative study of adolescent image-sharing practices on social media. *International Journal of Child-Computer Interaction, 20*, 64–71. https://doi.org/10.1016/j.ijcci.2019.03.002

Bergqvist, K. (2012). "Own work" in primary school: A teaching and learning practice in the context of administration and control. *Education Inquiry, 3*(2), 283–296. https://doi.org/10.3402/edui.v3i2.22034

Broady, D. (2007). Den dolda läroplanen. *Kritisk Utbildningstidskrift (KRUT), 127*(3), 13–98. https://www.skeptron.uu.se/broady/arkiv/dba-b-20070828-broady-den%20dolda-omtryck-krut-no-127-faksimil.pdf

Carlgren, I. (2015). *Kunskapskulturer och undervisningspraktiker*. Daidalos.

Dovemark, M. (2004). *Ansvar - flexibilitet - valfrihet: En etnografisk studie om en skola i förändring*. Acta Universitatis Gothoburgensis.

Erlandson, P., & Karlsson, M. R. (2018). From trust to control - The Swedish first teacher reform. *Teachers and Teaching, 24*(1), 22–36. https://doi.org/10.1080/13540602.2017.1379390

Erlandson, P., Strandler, O., & Karlsson, M. R. (2020). A fair game – The neoliberal (re)organisation of social and relational practices in local school settings. *British Journal of Sociology of Education, 41*(3), 410–425. https://doi.org/10.1080/01425692.2019.1707067

Goffman, E. (1990). *The presentation of self in everyday life*. Penguin.

Hadwin, A. F., Järvelä, S., & Miller, M. (2017). Self-regulation, co-regulation and shared regulation in collaborative learning environments. In D. Schunk & J. Greene (Eds.), *Handbook of self-regulation of learning and performance* (pp. 65–86). Routledge.

Heath, C., Hindmarsh, J., & Luff, P. (2010). *Video in qualitative research: Analysing social interaction in everyday life*. SAGE.

Larsson, S. (2005). Om kvalitet i kvalitativa studier. *Nordisk Pedagogik, 25*(1), 16–35. https://www.diva-portal.org/smash/get/diva2:245080/fulltext01

Lindgren, S. (2015). Unga fildelningskulturer. In S. Lindgren (Ed.), *Ungdomskulturer* (pp. 127–140). Gleerups.

Löfgren, H., Pérez Prieto, H., Tanner, M., Löfgren, R., Samuelsson, J., & Lindberg, V. (2017). Att ständigt bli bedömd. In H. Pérez Prieto & H. Löfgren (Eds.), *Att ständigt bli bedömd - elevers berättelser om betyg och nationella prov* (pp. 175–188). Studentlitteratur.

Organisation for Economic Co-operation and Development (OECD). (2016). *Definition and selection of competencies (DeSeCo)*. Retrieved March 31, 2024, from https://www.oecd.org/education/skills-beyond-school/definitionandselectionofcompetenciesdeseco.htm

Rönn, C. (2023a). *Backstage pedagogy: Compulsory school pupils' informal social strategies when dealing with formal individual writing assignments for assessment*. Doctoral dissertation, Linnaeus University. https://urn.kb.se/resolve?urn=urn:nbn:se:lnu:diva-125604

Rönn, C. (2023b). Students' social strategies in responding to leaked national tests at a Swedish municipal compulsory school. *Cogent Education, 10*(2), 2253711. https://doi.org/10.1080/2331186X.2023.2253711

Rönn, C. (2024). Pupil's informal social strategies in a Swedish compulsory school – What pupils do and say, out of sight of the teachers, while managing written individual assignments. *Educational Review, 76*(3), 605–627. https://doi.org/10.1080/00131911.2022.2054955

Rönn, C., & Pettersson, D. (2023). Swedish students' everyday school life and teachers' assessment dilemmas: Peer strategies for ameliorating schoolwork for assessment. *Educational Assessment, Evaluation and Accountability, 35*, 37–66. https://link.springer.com/article/10.1007/s11092-022-09400-3

Schunk, D. H., & Greene, J. A. (2018). Historical, contemporary, and future perspectives on self-regulated learning and performance. In D. H. Schunk & J. A. Greene (Eds.), *Handbook of self-regulation of learning and performance* (pp. 1–18). Routledge.

Sundberg, D. (2021). *Svenska läroplaner: Läroplansteori för de pedagogiska professionerna*. Studentlitteratur.

Swedish National Agency for Education. (2011). *Läroplan för grundskolan, förskoleklassen och fritidshemmet* [Curriculum for the compulsory school, preschool class, and school-age educare]. Skolverket.

Swedish National Agency for Education. (2017). *Kommentarmaterial till kursplanen i svenska 2011: Reviderad 2017* [Comment material to the course plan in Swedish 2011: Revised 2017]. Skolverket.

Swedish National Agency for Education. (2022). *Läroplan för grundskolan, förskoleklassen och fritidshemmet* [Curriculum for the compulsory school, preschool class, and school-age educare]. Skolverket. https://www.skolverket.se/publikationsserier/styrdokument/2024/curriculumforcompulsoryschoolpreschoolclassandschoolageeducarelgr22.4.7c808d9519080d9657c2.html?id=13128

Winstone, L., Mars, B., Haworth, C. M. A., & Kidger, J. (2022). Types of social media use and digital stress in early adolescence. *The Journal of Early Adolescence, 43*(3), 294–319. https://doi.org/10.1177/02724316221105560

Zimmerman, B. J. (1989). A social cognitive view of self-regulated academic learning. *Journal of Educational Psychology, 81*(3), 329–339. https://doi.org/10.1037/0022-0663.81.3.329

Chapter 7
University Policies on Academic Ethics and Integrity in Bulgaria

Irena Vassileva and Mariya Chankova

Abstract The study presents an analysis of existing university policies pertaining to academic ethics and integrity, teaching and learning practices, the transparency of applying those practices, as well as the responsibility of all parties concerned. The policies of 52 higher education institutions in Bulgaria, which are officially recognised by the National Evaluation and Accreditation Agency, are examined. This investigation is the first stage of a comprehensive state of the art review of plagiarism and academic misconduct-related issues in Bulgaria and draws from previous findings of non-uniform perceptions of what qualifies as plagiarism breaches.

The study follows the methodological steps outlined in an Australian study for locating and coding the relevant policy documentation. The analysis starts from the assumption that policy should be based upon the principles of transparency, respect, and responsibility and that these principles are directly related to the quality of academic output in both students and academics. The results demonstrate that the said principles are not reflected in the policies of Bulgarian institutions and their application, and it is therefore highly recommended that the documentation be revised accordingly.

Keywords Ethics codes · Academic ethics · Bulgaria · Academic integrity policy · Academic integrity

I. Vassileva (✉)
New Bulgarian University, Sofia, Bulgaria

M. Chankova
South-West University Neofit Rilski, Blagoevgrad, Bulgaria

I. Glendinning et al. (eds.), *Ethics and Integrity in Education (Research)*, Ethics and Integrity in Educational Contexts 9,
https://doi.org/10.1007/978-3-031-98157-9_7

Introduction

In a dynamic academic environment, different stakeholders—instructors, students, policymakers, administrators—have been facing numerous challenges, not least of which is articulated in terms of incorporating technological advances in the teaching and learning processes in a sustainable way, in harmony with the principles of academic integrity. The clear vision about these principles and the way they underscore the workings of an entire academic institution are reflected in policy, which is not merely a regulating or corrective tool, but should also provide the necessary bridge towards actual pedagogical practices oriented towards learning (for example, Bertram Gallant, 2017; Stoesz & Eaton, 2022). In the past decade or so, a critical volume of studies focused on policy evaluation has emerged (Anohina-Naumeca et al., 2020; Löfström et al., 2015; Miron et al., 2021; Stoesz & Eaton, 2022), articulating the idea that university policy not only adopts as guidelines the principles of academic integrity, but also explicitly translates those into concrete and applicable procedures, in which the responsibility of each participant is clearly articulated.

This contribution turns to Bulgarian university policies pertaining to academic ethics and integrity, teaching and learning practices, the transparency of applying those practices, as well as the responsibility of all parties concerned. The pilot investigation is the first stage of a comprehensive state of the art review of plagiarism-related issues in Bulgaria, and it draws from previous findings of non-uniform perceptions of what qualifies as academic integrity breaches (Vassileva & Chankova, 2019). The interest in university policy on academic integrity is without precedent in the Bulgarian context, with researchers typically focusing on individual cases of academic integrity breaches instead. Therefore, the present empirically driven analysis is expected to throw light on the current content of academic integrity policy documents in order to evaluate their reliability and practical applicability.

The underlying assumption for the present study is the following: an official university policy on academic integrity is a reflection of the way academic integrity is situated in the university's ecosystem and in teaching practices, thus being an integral part of the ethical foundation of young researchers and professionals. University policy regulating research and staff promotion in Bulgaria for both public and private universities has a common legal foundation provided by the Ministry of Education. Each university is free to apply the law within the legal framework provided. The issues of academic integrity and plagiarism are not objects of legal decision in the same way. What is more, in Bulgaria, there is no theoretical concept of academic ethics (and, by extension, academic integrity) which could become the basis for centralised policy and/or law; this fact was confirmed by the chairperson of the Ethics Committee in Bulgaria, during a round table discussion organised as part of the project. Thus, our focus in this study was directed towards current institutional policy documents that apply the concept of academic integrity (ethics) and refer to plagiarism and related topics.

In an overview of academic integrity policy and practices in EU institutions, Glendinning et al. (2013) rated Bulgaria 27th out of 27 countries surveyed, noting a culture of reluctance to discuss matters related to plagiarism, lack of research, a tendency to laying blame on teachers for the students' unsatisfactory performances, and an atmosphere in which "more enlightened approaches to discouraging or penalising dishonesty are discouraged" (p. 12). The report revealed that, for an important part of the EU countries surveyed, plagiarism and academic integrity did not appear to be prioritised at higher education institutions at the time, leaving individual researchers without the necessary support to advance in raising awareness about issues correlating with different facets of academic integrity, as policy was predominantly found to be inadequate. Several years later, looking into South-Eastern European institutions (Bulgaria was not among the investigated countries), Glendinning et al. (2017) pointed out that a prevalent attitude towards plagiarism offences was to ignore them, with a general lack of enforcing guidelines on academic integrity policies. The data collected from self-reports include information about codes of ethics and similar policy documentation, while at the same time, respondents admitted to ineffective top-down communication, coupled with a lack of evidence about the policy's implementation and a "relaxed attitude to cheating" (Glendinning et al., 2017, p. 136). It is interesting to note that the worst performing countries, according to Glendinning et al. (2013)—Bulgaria, Luxemburg, Italy, France, Spain, and Romania, are quite diverse in terms of their geographic location and societal tradition. Foltýnek and Glendinning (2015) pointed out differences in responses to questions from respondents in "East" and "West" European countries. Their study shows that students from western parts of Europe have a heightened awareness about matters related to plagiarism in comparison to respondents from eastern Europe. In addition, the study found that students in Western Europe were more likely to have received training about academic integrity and plagiarism than students in eastern parts of Europe (Foltýnek & Glendinning, 2015). However, a heightened awareness of plagiarism-related matters in one country is just one aspect of a complex set of interrelated factors about institutional policy, including prevention, detection, attitude, sanctions, and research.

The importance of academic integrity policy has been approached from multiple perspectives. Bretag et al. (2011) outline five core elements of the exemplary policy: access, approach, responsibility, detail, and support. These core elements have been since used to gauge the quality and the possibility for successful practical application of a coherent academic integrity policy (Anohina-Naumeca et al., 2020). Practical application is of great importance, especially in an integrated approach, implementing policy as part of teaching and learning (East, 2009). Some researchers have framed the issue from the students' perspective as a teaching and learning issue, which shifts the weight from detecting and correcting (or penalising) wrongdoing to fostering an environment which puts more value on learning and development (Bertram Gallant, 2017). A similar idea has been explored by Chankova (2020), connecting a knowledge-oriented learning environment, personal motivation, and cheating. The responsibility in terms of academic integrity does not only fall on individuals–although every individual is responsible for their own integrity–policy embodies the share of collective or institutional responsibility to uphold and foster principles that provide the environment for individual responsibility (Shephard et al., 2015). To sum up, the necessity for providing clear, detailed and applicable policies encompassing and safeguarding the interests of all parties concerned within the academic communities of educational and research institutions is of crucial importance for the maintenance of academic integrity.

The increasing attention towards developing academic integrity policy lies in its being identified as the foundation of all other aspects of education (Bretag, 2018), evolving mostly around honesty (Löfström et al., 2015; Bretag et al., 2011; Shephard et al., 2015, to name a few). According to the International Centre for Academic Integrity, the core values that embody the concept of academic integrity are honesty, trust, fairness, respect, responsibility, and courage, which should be upheld at all times (ICAI, 2014). Even though several documents pertaining to academic integrity in the EU were published in 2005, including the European Charter for Researchers and the Code of Conduct for the Recruitment of Researchers (European Commission, 2006), indicating a clear desire to lay a common values foundation in academia, they did not appear to have been incorporated in national policy or practices by 2012, as evidenced by Glendinning et al. (2013).

In Bulgaria, one of the landmarks in the development of a policy-based academic culture was the adoption of a new law regulating the recruitment and promotion of academic staff in 2010 (Ministry of Education and Science, 2010). This legislation replaced the former centralised Higher Accreditation Commission and gave the institutions autonomy in awarding academic titles. In addition, a national Ethics Commission was created in 2018, which is part of the Ministry of Education and Science. The new legislative base was adopted in each higher education institution, which resulted in the creation of a large body of documentation, some of which was dedicated to issues related to academic integrity. An amendment also required that all higher education institutions introduce a Code of Ethics as part of their formal documentation. It is these documents—policy and rule books—that we turned to in order to understand how academic integrity is perceived and applied in Bulgarian universities. We started from the assumption that transparency, respect and responsibility were minimally reflective of the core academic principles, in that they involve the foundation of both institutional and personal participation in the furthering of academic culture.

Aims and Method

Our aim was to establish the principles around which institutional policies are currently formulated and how they relate to the organisation of teaching and research at Bulgarian universities. The following research questions formed the focus of this investigation. First, what categories underlie the academic policy standards at Bulgarian higher education institutions? Second, how do institutional policies align with centralised efforts to streamline issues of academic integrity? And third, how do these policies relate to teaching practices and staff support? The results of the qualitative analysis were mapped onto Ministry recommendations in order to acquire a comprehensive picture of the state of institutional attitudes towards academic integrity issues.

For our study, we examined the websites of all 52 higher education institutions (both state-owned and private) recognised by the National Evaluation and Accreditation Agency at the time of gathering the data (between February and April 2023) to collect the relevant documentation. Out of the 52 institutions, 28 had a publicly accessible Code of Ethics or a related document explicitly identifying issues of academic integrity that could be found online. The documents were not easy to locate, and in all the cases, the website's integrated search engine had to be used to retrieve the policy. The accessibility and ease of locating the policies were noted as these factors directly connect to an important criterion for exemplary policy identified by Bretag et al. (2011). Along with the implementation date and the title of the policy, these features are described as the formal aspects of the documents. The documents were then qualitatively examined by both researchers separately, and the observations and notes were pooled.

The following qualitative essential aspects of analysis were identified:

- The underlying principles or values, either explicitly or implicitly identified;
- The presence of definitions of fundamental notions, such as plagiarism, academic integrity, or various types of breaches of academic integrity;
- The procedures for reporting suspected violations and provisions for relevant sanctions;
- The target audience of the policy and the clear assignment of the responsibility.

The criteria enumerated above closely follow those determined by Bretag et al. (2011) and were considered applicable to the Bulgarian reality. Besides, their analysis could serve as a basis for facilitating the creation and introduction of appropriate policies in Bulgaria, aligned with international practices. They could also be useful in future comparative studies with policies in other countries and institutions to elicit good practices.

In order to gain perspectives and understand the attitudes of various stakeholders, in April 2023 a special Round Table was held under the topic "Plagiarism. Methodological problems; interdisciplinary perspectives". In attendance, members of the Ministry of Education, the Ethics Commission, lawyers, publishers, university professors, and policy-makers expressed their perspectives in a non-structured, freeform discussion. These discussions provided impetus for the further development of the project, including a much-needed background for the interpretation of our findings.

Results and Analysis

Formal Aspects of the Documents

As mentioned above, in terms of "Access to policy," in all cases the respective documents were "well-hidden" and practically impossible to find without using the search function on the respective university website, no matter who tried to locate them. The years of implementation/adoption of the policies are as follows:

- 19 policy documents were adopted before 2018;
- 5 policy documents were adopted after 2018;
- 4 documents were undated.

All policies follow the European Charter for Researchers (proposed by the European Commission in 2005) very closely and cite it. Most policies have the title "Code of ethics" and often refer to other documents regulating the established university Ethics Committees. Many of them also contain in their title the phrase "академично единство" ("Academic unity")—a term which is not used in any of the existing official legal documents, and neither is it defined or explained anywhere in the respective texts. The term could be a translation from the English phrase "academic unity," used in academia to designate a vision of academia as sharing common goals in the creation and synthesis of knowledge (as used by Miller, 2016), which is itself an interpretation of the notion of "unity of science". Frank and Meyer (2020), on their part, envisage 'unity' as an organisation-internal feature, arguing that university study programs are globally similar in their aspirations despite local socioeconomic differences. They attribute this to the enduring, integrated structure of universities, where faculties and schools form a unified whole. This results in global isomorphism in academic content, though variations in organisational structures ensure that cultural content remains adaptable and open to interpretation.

Notably, in the investigated Bulgarian documents, the term "academic unity" appears only in the titles ("Code of Ethics and Academic Unity") and in the designations of the established committees. It does not seem to correlate to any of the above understandings, since the overall spirit of the texts suggests an interpretation of the term as primarily calling for loyalty to the institution and, in practice, often leads to breaches of academic integrity being ignored. Attempts to put an end to this destructive practice have been provided in a recent amendment to the Higher Education Law, postulating that, in cases of proven plagiarism by a staff member, if the rector of the university concerned does not take the respective measures required by the law (usually to expel the plagiarist), they will lose their position.

Essential Aspects

Purpose of the Policies

Succinctly summarised, the stated purpose of the policies is to provide guidelines for ethical behaviour, help promote the academic and pedagogical mission of the institution and maintain its positive image in society. Many policies enumerated principles which must be adhered to, namely: legality, loyalty, honesty, impartiality, competence, responsibility, accountability, tolerance, and respect for the individual. They also aimed to define and regulate professional behaviour, collegiality, conflict of interest, academic freedom and responsibility, civil responsibility, and loyalty to the institution, among others.

Key Terms and their Definitions

Only six documents provided something of a definition of "Academic integrity," such as:

"A set of specific ethical principles to regulate staff behaviour," "science and truth, transparency, authorship, veracity," "a kind of professional ethics reflecting the particularities of the institution."

The rest relied on lists of ethical values without any clarifications of what they represent.

Ten documents had no definition of "plagiarism," 18 documents contained elements of a definition, and one referred to the Bulgarian Penal Code as the source of definition, for instance:

> Publishing the work of somebody else entirely or partly as one's own, incorrect use of somebody else's ideas or projects, scientific results or their interpretation, especially when such an action leads to undeserved gains and to harming the image of the university.

In addition, none of the documents provided links to any additional sources that may have contained explanations or definitions. There were no mentions of plagiarism detection tools and/or how they could be implemented either as pre-emptive instruments or for checking suspicious cases. It was only in 2023 that the Ministry of Education and Science acquired plagiarism detection software and put it at all universities' disposal for free. However, in spite of the country-wide campaign for promoting the software and training its potential users, academic staff at some of the universities are reluctant to employ it (highlighted during the Round Table discussions). It could be speculated that the reasons identified in Vassileva and Chankova (2019)—namely, the complexity of using the software, the time-consuming nature of plagiarism checks, and the lack of perceived value among academics and faculty in using text-matching software to assess students' work—may still apply.

As a general rule, authors of PhD theses sign and submit a declaration stating that the work is "their own," while at some universities, there is an additional procedure of submitting manuscripts to text-matching software at the discussion stage

Table 7.1 Academic Misconduct

Cases of academic misconduct	Number of documents
Undue use / credit	9
Proper citation	5
Falsification	4
Fabrication	3
Intellectual property	3
Copyright	2
Authorship	1
Pseudo-scientific work	1
Publishing same work under different title	1

(internal review before publication or viva). A procedural lack of clarity exists in relation to undergraduates' and post-graduates' (MA) academic production. The members of the committees awarding academic degrees and titles are obliged by law to scrutinise the materials. Unfortunately, they are, more often than not, disinclined to do so either due to lack of time or because they are unwilling to find themselves in conflict situations with other colleagues involved in such procedures. In other words, interpersonal relations in the small Bulgarian academic community play a major role in scholars' decisions on how to act in "uncomfortable" situations, therefore, these confrontations are avoided if possible (Vassileva & Chankova, 2019).

Moving on to the notion of classifying types of "Academic misconduct," Table 7.1 presents a list of its various types and the number of documents these appear in.

As the findings suggest, almost half of the terms included in the documents relate to plagiarism, with other types of misconduct receiving very little attention. At the same time, cases of results fabrication and/or falsification, as well as of pseudo-scientific work and problematic authorship, are constantly increasing worldwide and have become a major issue for publishers who are under pressure to retract numerous articles (see Retraction Watch and the Committee on Publication Ethics (COPE) for more statistical information and examples). Collusion, contract cheating and self-plagiarism are not included in any of the policy documents, although these forms of academic misconduct are very widely discussed globally. Collusion and contract cheating are common types of student misconduct, while self-plagiarism can apply to both students and academics. More specifically, previous research (Vassileva & Chankova, 2019) demonstrated that in Bulgaria, self-plagiarism is not generally perceived as a type of misconduct.

Five of the documents contained mention of "breach of policy," where one document relied on a distinction between "moral breach" and "breach of labour law" without any detailed explication. Another document outlined a general provision: "intellectual theft, fabricating and falsifying results". A further document stipulated

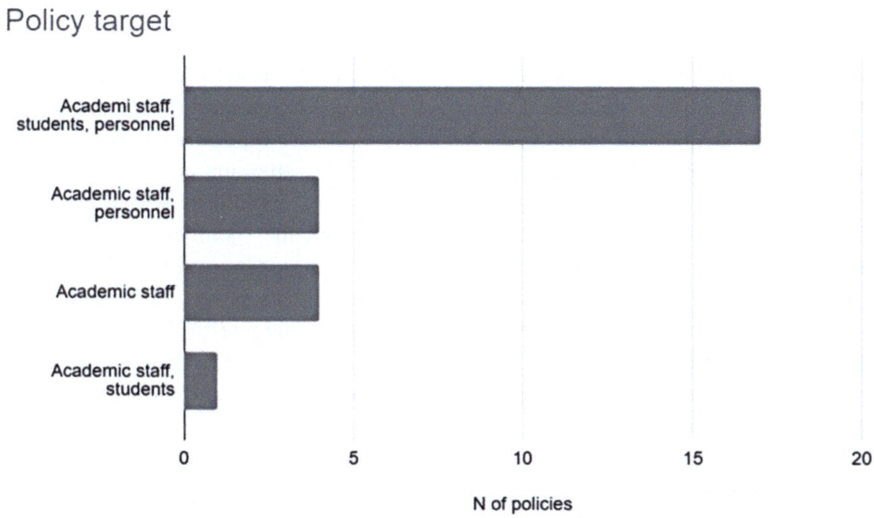

Fig. 7.1 Target of Policy

that a breach is "amoral behaviour which harms the university image". The rest of the documents generalised misconduct as: "Any deviation from the norms described in the Ethics Code". Considering the vague nature of the definitions (where there are definitions), it is the university Ethics Committee that decides whether what has been reported is a breach or not. Thus, the procedure as it is leaves room for individual interpretations and could result in inconsistencies and misjudgements leading to either unfairly harming someone's reputation or failing to penalise a clear breach of integrity.

In terms of the "Target of policy," Fig. 7.1 demonstrates to whom the respective policies are directed and, respectively, who is supposed to comply with them.

As shown in Fig. 7.1, most of the Codes we analysed included all university members. However, some Codes still exclude students or other personnel, which appears inadequate, given that students are generally considered the group most likely to violate these policies. This issue has been a longstanding concern, with numerous publications addressing cases of misconduct and exploring strategies for prevention and mitigation. The legislation requires that new appointees (academic staff) be notified of the Ethics Code when they start their employment. However, there is no explicit mention of how students are supposed to get informed of the policy. According to the Implementation Plan of the Strategy for the Development of Higher Education in the Republic of Bulgaria for the period 2021–2030 (Ministry of Education and Science, 2021), Provision 2.1.6. stipulates that all first-year students take a compulsory course on Academic Ethics to be introduced by 2023. How this provision has been realised in practice is still an open question.

Reporting Suspected Violations

According to the examined documents, suspected violations are reported to the respective university's Academic Ethics Committee. These committees consist of 5 to 9 members, depending on how large the university is. Typically, apart from academic staff members, there are representatives of the students and non-academic staff. The members are appointed for a term of 4 years, coinciding with the rector's term. An issue of concern, however, is the fact that the members of the Ethics Committees are selected by the rector or by the Academic Council after the rector's suggestion, as stated in most documents. The committees have an advisory function, and their conclusions are unbinding. They may suggest to the rector, in the capacity of the official employer, measures to be taken in cases of proven academic misconduct. However, at some universities, the rector has the authority to overrule, reverse and annul decisions of the Ethics Committee. For instance, one of the policies includes the following special provision:

"Art.24. The Rector, respectively the Academic Council of the University may cancel, modify and/or return for reconsideration the appealed decisions of the Ethics and Academic Unity Committee".

Thus, the selection procedures of the committees and the special provisions endow the rectors with the power to exert strong influence and potentially undermine the work of the committees, which might create mutual dependencies.

In contrast, the Ethics Commission, created by the Ministry of Education and Science in 2018, is tasked with dealing with contentious cases unresolved by the university Ethics committees, and its decisions are final and binding. The Commission reports directly to the Minister who orders the rectors to take the respective administrative measures anchored in the law (revoking the academic degree and dismissing the offender). If they refuse to do that, they run the risk of being dismissed themselves.

The "types of individuals/entities that can report academic misconduct" are shown in Fig. 7.2 below.

In the majority of documents in the study, there is no specific mention of who can report suspected violations, which may result in arbitrarily rejecting signals on the basis of not coming from an authorised person. Besides, some policies stipulate that anonymous signals should not be considered at all, which, in view of the abovementioned small size of the academic community and the closely-knit relationships and dependencies among its members, could lead to reluctance to report due to fear of retribution. Three policies insist on the confidential treatment of the signals, while one states that the records of the Ethical Committee are freely accessible by all staff members. It is unclear whether the identity of the reporting individual is kept confidential, at least for the duration of the Committee's investigation. The lack of transparency regarding these matters effectively prevents any principled approach towards investigating violations.

Moreover, in contrast to similar documents in other countries, there are no provisions granting the rights and safety of the person reporting allegations of misconduct, such as: "The University promises that there will be no adverse action,

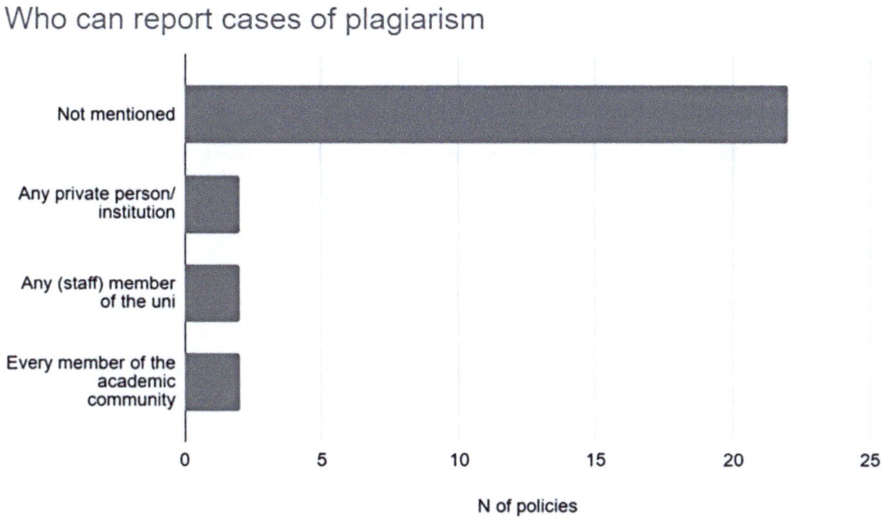

Fig. 7.2 Who Can Report Cases of Academic Misconduct?

retribution, or other reprisal for the good faith reporting of a suspected violation of this Code, even if the allegations ultimately prove to be without merit." (NYU). Neither are there stipulations that: "Each member of the University is expected to uphold the standards of New York University and to report suspected violations of the Code or any other apparent irregularity" (NYU). Two policies also contain the addition that reports must be accompanied by proof without specifying what counts as proof. As a result of these flaws, the parties involved may feel no obligation to report, prefer to remain silent or may fear retaliation for doing so. The latter is somewhat understandable as, in fact, there have been several cases of university authorities turning against whistleblowers, blaming them for blemishing the image of the university instead of penalising the real violators, some cases being widely relayed by the media (for example, Georgieva, 2024; Georgieva, 2025).

Therefore, it is no surprise that almost all policies place special focus on members' maintaining "the good name" of the university, refraining from information leaks to the general public, which may put the institution in a bad light, even in the event of serious breaches of academic ethics. The following examples from the data pool illustrate this general trend, without representing an exhaustive list (translations have been done by the authors):

> "It is not permissible for lecturers and employees to spread information about irregularities in the [university] in print and electronic media, on online social networks or in any other way".

> "Employees are obliged to keep secret data, facts and circumstances, the disclosure of which may harm the interests of the University, its good name, or damage its relations with third parties".

> "In the performance of their duties and in public life, University employees shall behave in a manner that does not bring the University into disrepute".

> "Members of the academic community of the University must be loyal to the University and not bring the University into disrepute among the public".

> "Confidentiality. Members of the academic community and employees shall protect and not disclose the facts and circumstances which come to their knowledge in the course of or in connection with the performance of their duties".

In some cases, the provisions are highly restrictive:

> "Only persons authorized for this purpose by the Rector's Body or the Dean's Body (PR) may communicate with the media".

To sum up, the unclear and non-transparent reporting procedure, the misinterpretation of loyalty to the institution, and the existing dependencies within Bulgaria's relatively small academic community result in reluctance to report cases of plagiarism or any violations of academic integrity.

Sanctions

Nineteen documents provide directions relating to sanctions, where 12 refer to disciplinary sanctions outlined in the Bulgarian Labor Code, two mention "administrative sanctions" enforced by the Rector, and five suggest "moral sanctions" (see below for definitions). However, in fact, the sanctions are described by the Act on the Development of Academic Staff in the Republic of Bulgaria and the Higher Education Law—both of which are not mentioned in the Labor Code at all. Thus, it seems that the three legal documents have not been aligned to cater for clear instructions for the implementation of sanctions related to academic misconduct. In addition, it is important to note that many policies explicitly stipulate that the Ethics Committees are not mandated to enforce sanctions. The mandate of the Ethics Committee appears to be limited to ruling on breaches, and then it is for the rector to decide whether the breach calls for sanctions under the provisions of Labor Law. One policy indicates that the Committee's recommendations are applicable on a voluntary basis.

Moral sanctions were defined in one document as: "Critical publication in the scientific press, mass media, social networks, and other public outlets can serve as negative moral judgment conduits, relaying moral condemnation after the violation of ethical norms has been unequivocally established". The disciplining effect of public shaming (as a kind of moral sanction) has been identified as a possible route towards changing public attitudes vis-a-vis plagiarism. The need to foster a culture of non-tolerance towards plagiarism as a breach of academic integrity was also articulated by other academics (Round Table Discussions).

The phrase "administrative sanctions" is not explained in the policies in any way. The lack of a legal definition for the term "academic ethics" (Daskalov, 2013; Round Table discussions) may be the reason for the lack of specified sanctions in the documents, reflecting the particularities which legal researcher Daskalov identifies for academic ethics: "In general, the rules representing in their system academic ethics, do not have a legally binding character [...] For this reason, the sanctions for breaches of the ethical rules usually develop in the moral sphere – i.e., the violation of the moral and ethical rules, normally, does not lead to any negative legal consequences such as the imposition of penalties, but are related to damage to prestige and good name" (Daskalov, 2013, p. 16). It should also be mentioned that in the Bulgarian legal system, plagiarism has never been conceptualised as a breach of academic ethics (Round Table discussions, Chairman of the Ethics Commission).

Conclusions

To begin with, the problematic access to the policies demonstrated a lack of transparency and insufficient recognition of the importance of academic integrity. The Ethics Codes examined claimed to adopt academic integrity as a leading principle but failed to provide clear and full definitions of what this constitutes. Simplified and vague (if any) definitions of plagiarism and other breaches of academic integrity were included, while some serious types of misconduct were not considered at all. There is no mention of tools to help detect plagiarism, although such a tool was provided for free by the Ministry of Education and Science in 2023. This may partly be due to the fact that very few policies state the option of being open to revision, and the latest revisions are from 2018. Even conceding the latter point, student support is the big omission from the policies, which do not explicitly articulate in which way students are required to uphold the ethical guidelines and what these are. The conclusions are summarised in Table 7.2.

The parties/stakeholders that are allowed to report allegations of academic integrity breaches remain largely unspecified, which may lead to the rejection of allegations. The small size of academic communities in Bulgaria, where practically everyone knows everyone else in a specific field, combined with close personal relations and dependencies, results in a reluctance to raise allegations. The lack of guarantees of protection against retaliation for individuals who raise allegations, the vagueness of envisaged penalties and the almost unlimited powers of rectors to overturn decisions made by ethical committees are additional factors that deter scholars from reporting allegations of violations.

In the documents examined, sanctions were mainly covered by referring to other legislative acts where the respective laws are not aligned, leaving space for various interpretations at different institutions, creating opportunities for unequal treatment and discrimination across the academic community and the country. The so-called "moral sanctions," on their part, such as "naming and shaming", are usually quickly forgotten and regarded as "a nine days' wonder." Only cases of academic misconduct involving outstanding public and political figures that provoke the attention of the mass media may remain in the public realm for longer and may eventually lead to legal sanctions.

Table 7.2 A Summary of the Findings in Accordance with the Core Elements of the Exemplary Policy on Academic Integrity (following Bretag et al., 2011)

Core elements of academic integrity (Bretag et al., 2011)	Our findings based on 28 policy documents
Access	1. Not all institutions made policies available on their websites. 2. When policies were available, they were not highlighted on the university websites; a search engine was necessary to locate them.
Approach	1. The policies established an ethics committee—a cross-institution group in charge of investigating breaches of policy. 2. The mandate of the committee was limited and often non-binding.
Responsibility	1. The committee's responsibilities were described. 2. Some policies referred in general terms to the staff, personnel and students being held to uphold the ethics code.
Detail	1. Some policies contain general statements about reporting breaches. 2. No specific policy or procedures were identified as to how to apply the provisions in the ethics code. 3. Anonymous reporting was excluded. 4. No penalties were specified other than the provisions in the labor code (which do not deal specifically with academic misconduct) and moral sanctions. 5. No clarifications on terminology, definitions and examples were provided.
Support	1. No provisions on supporting staff or students in mastering and applying the principles of academic integrity were provided. 2. No strategies for engaging staff in promoting and fostering a culture of academic integrity were provided.

Last but not least, the investigated documents did not provide evidence of any kind of support to promote understanding and implementation of the policies' principles.

To sum up, the Higher Education Law requires the members of the academic community to comply with the standards of scientific and professional ethics, which has led to the creation of Codes of Ethics as part of the internal regulations of the universities. However, as Nikolova (2019) duly noted, there are no "standardised norms of ethical behaviour and procedures for solving ethical problems developed by experts, anchored in a law and applied throughout the whole higher education system" (p. 73) in Bulgaria. As a consequence, disparate approaches and practices are observed, which result in unequal treatment of similar cases of academic misconduct. This, paired with personal dependencies, often becomes a prerequisite for some scholars to assume that "anything goes" or, for others, to feel as the victims of a dysfunctional system and resort to resignation.

Regulations on the use of generative AI, such as ChatGPT and similar, are non-existent as yet in Bulgaria, and no serious discussions have been started along these lines at scholarly or administrative levels at higher education institutions at the time of writing this chapter. At the same time, the European Commission AI Act (EU AI

Act) states that: "AI systems identified as high-risk include AI technology used in: […] educational or vocational training, that may determine the access to education and professional course of someone's life (e.g., scoring of exams) […]" (European Commission, 2023). Universities worldwide have started to develop and implement policies on the use of AI, although the approaches vary from totally prohibiting such use to supporting students and staff to become AI-literate and equipping them with the knowledge on how to ethically apply generative AI tools.

Recommendations for Improvement

Providing detailed and full definitions of all terminology relating to academic integrity will disambiguate existing misconceptions and inconsistencies and offer a clear picture of what is needed.

Providing comprehensive and explicit definitions of all types of breaches of academic integrity and articulating sanctions or penalties that are fair and appropriate will make stakeholders aware of the seriousness of the problems and the expected consequences.

Providing a list of stakeholders, their rights to report allegations of academic misconduct and the measures in place to protect them will encourage reporting of allegations.

The development of support systems and instruments, especially directed towards students and graduates, including consistent utilisation of text-matching software, will help to educate all stakeholders about accidental plagiarism, deter against deliberate plagiarism and help to better align policy with teaching practice (East, 2009). This provision will especially benefit academics and researchers, especially early career academics, to support their engagement and development.

The gradual development of policies on AI use, in line with EU policies and regulations, will help to restrain students and academics from abusing the new technological tools for further fraudulent advancement in their studies and research. Instead, the policy should stipulate rules and measures on how to ethically employ the technology for innovative approaches to teaching, learning and research.

Finally, providing easy and transparent access to Ethics Codes and supporting documents on university websites will facilitate all interested parties in becoming acquainted with them and highlight the importance of developing a culture of academic integrity for participation in academic life in general.

In a nutshell, this study demonstrates that there are numerous problems with the setting of standards of academic integrity in Bulgaria, and more multifaceted research is needed encompassing various fields such as psychology, law, educational policy, and information technology.

Acknowledgments This research is supported by the Bulgarian National Science Fund under grant number КП-06-Н70/9 (2022-2025).

References

Anohina-Naumeca, A., Birzniece, I., & Odiņeca, T. (2020). Students' awareness of the academic integrity policy at a Latvian university. *International Journal for Educational Integrity, 16*(12), 1–17. https://doi.org/10.1007/s40979-020-00064-4

Bertram Gallant, T. (2017). Academic integrity as a teaching and learning issue: From theory to practice. *Theory Into Practice, 56*(2), 88–94. https://doi.org/10.1080/00405841.2017.1308173

Bretag, T. (2018). *Academic integrity*. Oxford Research Encyclopedia of Business and Management. https://doi.org/10.1093/acrefore/9780190224851.013.147

Bretag, T., Mahmud, S., Wallace, M., Walker, R., James, C., Green, M., East, J., McGowan, U., & Partridge, L. (2011). Core elements of exemplary academic integrity policy in Australian higher education. *International Journal for Educational Integrity, 7*(2), 3–12. https://doi.org/10.21913/IJEI.v7i2.759

Chankova, M. (2020). Teaching academic integrity: The missing link. *Journal of Academic Ethics, 18*(2), 155–173. https://doi.org/10.1007/s10805-019-09356-y

Committee on Publication Ethics (COPE). (n.d.). Committee on Publication Ethics. https://publicationethics.org/

Daskalov, V. (2013). Академичната етика и образованието по право [Academic ethics and law education]. https://www.unwe.bg/uploads/ResearchPapers/Research%20Papers_vol2_2013_No1_V%20Daskalov.pdf

East, J. (2009). Aligning policy and practice: An approach to integrating academic integrity. *Journal of Academic Language and Learning, 3*(1), 38–51. https://journal.aall.org.au/index.php/jall/article/view/66

European Commission. (2006). *The European charter for researchers – The code of conduct for the recruitment of researchers*. https://op.europa.eu/publication-detail/-/publication/d02f13fa-4bb8-4f33-90ee-fc20df447cfc.

European Commission. (2023). *Artificial intelligence act*. https://digital-strategy.ec.europa.eu/en/policies/regulatory-framework-ai.

Foltýnek, T., & Glendinning, I. (2015). Impact of policies for plagiarism in higher education across Europe: Results of the project. *Acta Universitatis Agriculturae et Silviculturae Mendelianae Brunensis, 63*(1), 207–216. https://doi.org/10.11118/actaun201563010207

Frank, D., & Meyer, J. (2020). *The university and the global knowledge society*. Princeton University Press.

Georgieva, S. (2024, September 24). *"Почти 100% чужди текстове": Етичната комисия получи сигнал за плагиатство на ректор* ["Almost 100% plagiarized texts": The Ethical Commission received a report for plagiarism violation against a rector]. Dnevnik. https://www.dnevnik.bg/bulgaria/2024/09/21/4673061_pochti_100_chujdi_tekstove_etichnata_komisiia_poluchi/

Georgieva, S. (2025, January 3). *МОН и борбата с плагиатството: надежда всяка тука оставет* [Ministry of Education and Science and the fight against plagiarism: Abandon all hope, ye who enter here]. Sega. https://www.segabg.com/category-observer/mon-i-borbata-plagiatstvoto-nadezhda-tuka-vsyaka-ostavete

Glendinning, I., Foltýnek, T., Demoliou, C., Joswisk, K., & Stabingis, L. (2013). Comparison of policies for academic integrity in higher education across the European Union. *IPPHEAE Project Consortium*. http://plagiarism.cz/ippheae/files/D2-3-00%20EU%20IPPHEAE%20CU%20Survey%20EU-wide%20report.pdf

Glendinning, I., Foltýnek, T., Dlabolová, D., Linkeschová, D., & Lancaster, T. (2017). South East European project on policies for academic integrity. *IPPHEAE Project Consortium*. http://www.plagiarism.cz/seeppai/Final-report_SEEPPAI.pdf

International Center for Academic Integrity (ICAI). (2014). *Fundamental values*. https://academicintegrity.org/aws/ICAI/pt/sp/values

Löfström, E., Trotman, T., Furnari, M., & Shephard, K. (2015). Who teaches academic integrity and how do they teach it? *Higher Education, 69*, 435–448. https://doi.org/10.1007/s10734-014-9784-3

Miller, M. J. (2016). Academic unity: The creation and synthesis of knowledge. *Graduate Research Journal*, 36–49.

Ministry of Education and Science. (2010). Закон за развитие на академичния състав на Република България [Promotion of academic staff act. normative acts]. https://www.mon.bg/nfs/2023/01/zkn_akademichnia-sastav_akt03102023.pdf

Ministry of Education and Science. (2021). *Implementation plan of the strategy for the development of higher education in the Republic of Bulgaria for the period 2021–2030*. www.mon.bg/mon/strategii-i-politiki/strategicheski-dokumenti-2/

Miron, J., McKenzie, A., Eaton, S. E., Stoesz, B., Thacker, E., Devereaux, L., Perasud, N., Steeves, M., & Rowbotham, K. (2021). Academic integrity policy analysis of publicly-funded universities in Ontario, Canada: A focus on contract cheating. *Canadian Journal of Educational Administration and Policy, 197*, 62–75. https://doi.org/10.7202/1083333ar

New York University (NYU). (n.d.). Code of ethical conduct. https://www.nyu.edu/about/policies-guidelines-compliance/policies-and-guidelines/code-of-ethical-conduct.html#reportingsuspected

Nikolova, B. (2019). Етичността в сферата на висшето образование [Ethics in the field of higher education]. *Диалог, 2*, 72–111.

Retraction Watch. (n.d.). Retraction watch. https://retractionwatch.com/

Shephard, K., Trotman, T., Furnari, M., & Löfström, E. (2015). Teaching research integrity in higher education: Policy and strategy. *Journal of Higher Education Policy and Management, 37*(6), 615–632. https://doi.org/10.1080/1360080X.2015.1102823

Stoesz, B. M., & Eaton, S. E. (2022). Academic integrity policies of publicly funded universities in western Canada. *Educational Policy, 36*(6), 1529–1548. https://doi.org/10.1177/0895904820983032

Vassileva, I., & Chankova, M. (2019). Attitudes towards plagiarism in academia. *English Studies at NBU, 5*(1), 135–163. https://doi.org/10.33919/esnbu.19.1.7

Chapter 8
Towards an Ethical Standard: An Analysis of Perspectives and Experiences in Complying with Basic Ethical Requirements

Shiva Das Sivasubramaniam

Abstract Research involving human subjects requires adherence to ethical guidelines that are dictated by funding agencies' requirements and adjudicated by ethical review boards. Despite strict requirements, ethical standards are not always being met. Here, we present cases that exemplify positive and negative activities related to ethical research conduct, as well as factors that influence these activities, in order to demonstrate that current policies and procedures related to implementing ethical research are insufficient and need to be improved.

Narrative analysis was carried out amongst researchers from multi-disciplinary subject areas (STEM and non-STEM). The data was collected through semi-structured interviews conducted between 2021 to 2024. The study included graduate students (GS – n = 10), postdoctoral researchers (PDR – n = 10), and research supervisors (RS – n = 15). The interviews are based on participants' experiences in conducting research and their interactions and experiences with peers and collaborators in this context.

The results show that participants carrying out externally funded research are adhering to ethical standards and expectations of their funding bodies and associated governmental organisations. There is evidence for institutional representatives carrying out monthly audits on procedural updates and health and safety. However, claims of unethical and/or immoral behaviours or practices (such as selective reporting of data, hypothesising after the results are known (HARKing), duplicate publications) were reported by the participants. Participant consensus indicated that institutions have a responsibility to establish and implement policies and procedures to support ethical research practice. Recommendations are provided on how institutions can minimise questionable practices and misconduct in research.

S. D. Sivasubramaniam (✉)
School of Life and Health Sciences, University of Roehampton, London, UK
e-mail: s.sivasubramaniam@roehampton.ac.uk

I. Glendinning et al. (eds.), *Ethics and Integrity in Education (Research)*, Ethics and Integrity in Educational Contexts 9,
https://doi.org/10.1007/978-3-031-98157-9_8

Keywords Ethics · Ethical behaviour · Research · Conduct · Standards · Questionable research practices

Introduction

Establishing acceptable ethical behaviours in research is crucial for maintaining integrity, safeguarding participants, and fostering public trust. Understanding compliance to basic ethical principles, such as honesty, objectivity, respect, confidentiality, and professional behaviour, in both STEM and non-STEM disciplines involves examining how researchers adhere to fundamental ethical guidelines across these disciplines. In STEM research, ethical guidelines focus on the responsible conduct of studies involving human subjects, their tissues, and the data derived from them. This includes informed consent, privacy protection, and minimising harm. Likewise, non-STEM research, encompassing the humanities and social sciences, also emphasises ethical considerations, particularly regarding the respectful treatment of participants and the responsible reporting of findings. By adhering to these ethical standards and by establishing ethical behaviours, researchers across all disciplines can contribute to the advancement of knowledge while upholding the highest moral principles.

In all cases, adherence to basic ethical principles is essential to maintain accountability and minimise risks (such as physical, psychological, and/or emotional harm to participants), resulting in loss of trust leading to legal challenges and reputational damage to the individuals (researchers) as well as the institutions (Snezhko & Coskun, 2019; National Academies of Sciences, Engineering, and Medicine, 2017).

"Ethical research conduct" does not mean simply following guidelines given in the respective ethical approval. It is a commitment to behaving ethically, respecting fellow researchers, research participants and health and safety rules and guidelines, in other words, establishing behaviour or culture that aligns with basic ethical principles. Acting this way provides a framework for public trust in the research findings (Sivasubramaniam et al., 2021; 2021; Mandal et al., 2011).

Researchers obtain appropriate ethical approval for their projects, not simply because it is a requirement, but to show their commitment to ethical research (CIOMS, 2017). As professionals, the scientific community is often expected to "self-regulate" and report any unexpected mishaps. However, due to internal and external pressures, often relating to establishing research esteem, researchers are being coerced to become egocentric, prioritising individual success and failing to follow the basic guidelines of ethical research as a result (Sivasubramaniam et al., 2021). Since the beginning of this century, cases of blatant scientific misconduct have received significant attention. For example, questionable research practices (QRPs)—behaviours and practices that, while not outright fraudulent, compromise the integrity, reliability, and validity of research—have been exposed by whistleblowers within the scientific community and publicised by the media (Altman,

2006; John et al., 2012). There are many reasons for QRPs. Some of the reasons for QRPs are discussed next.

Lack of reproducibility in research is often considered a symptom of poor scientific practices and, in some cases, research misconduct, rather than a direct cause of it. Studies that fail to reproduce may result from methodological flaws, inadequate reporting, or variability in experimental conditions rather than deliberate misconduct (Nelson et al., 2021; Baker, 2016).

However, on the other hand, persistent issues with reproducibility can erode confidence in research findings, potentially creating pressure on researchers to manipulate data, selectively report results, or engage in questionable research practices to align with expected outcomes (Ioannidis, 2005; Munafò et al., 2017). Such practices undermine scientific integrity, weaken public trust in research, and may lead to serious consequences, particularly in fields like medicine and public health, where unreliable findings can misguide policy decisions and clinical practices.

Likewise, institutional pressures that prioritise publication quantity over research quality also foster an environment that can encourage unethical practices such as the "publish or perish culture" (Rawat & Meena, 2014; Parchomovsky, 2000). Universities and funding bodies often use publication metrics, such as impact factors and citation count, as key performance indicators for hiring, promotions, and grant allocations. This creates a competitive, high-stakes atmosphere where researchers may feel compelled to rush experiments, cut corners, or even commit research misconduct to meet institutional expectations. The pressure to continuously produce novel findings also discourages replication studies and reporting of negative results, contributing to the reproducibility crisis. Ultimately, this system compromises research integrity, distorts scientific priorities, and may lead to burnout and mental health issues among academics, highlighting the urgent need for reforms that value rigorous, ethical, and reproducible research over sheer publication volume. The true extent of unreported scientific misconduct is challenging to quantify due to its concealed nature.

Surveys and studies are good ways to provide a glimpse into the problem. In 2008, a survey on research misconduct was carried out, where the participants were researchers funded by the United States National Institutes of Health (NIH). The results showed an alarming number of cases of potential research misconduct (Titus et al., 2008). Likewise, a meta-analysis carried out by Fanelli (2009) reported that 2% of scientists admitted to having committed fabrication, falsification, or plagiarism, while approximately 34% admitted to engaging in QRPs. Since people are unlikely to openly confess to misconduct, the reported cases or figures of misconduct may be probably lower than the actual number. In other words, the real scope of the problem is likely larger than what is revealed since many instances of misconduct might go unreported or undisclosed.

Therefore, the purpose of the study described in this chapter was to explore researchers' attitudes and their experience in (a) adhering to the basic ethical principles and (b) establishing ethical behaviours in every step of their research. We wanted to understand how many researchers are committed to abiding by basic ethical principles.

Historical Lessons: A Brief Literature Survey

The literature on reported high profile unethical medical research practices highlights cases as far back as 1932 (see Table 8.1 for summary). Three examples from this list will now be discussed to demonstrate how these cases influenced the development of the ethical principles in use today.

The Tuskegee Syphilis Study was a notorious unethical medical experiment conducted by the U.S. Public Health Service to study the progression of untreated syphilis in African American men (Brandt, 1978). Researchers misled participants by telling them they were receiving free medical care, while they were deliberately left untreated, despite even penicillin became the standard cure for syphilis in the 1940s. The study continued for 40 years, despite growing ethical concerns, until it was exposed by the media in 1972. This revelation led to widespread public outrage and significant reforms in research ethics, providing additional evidence for the establishment of the Belmont Report (Nagai et al., 2022) and stricter guidelines for informed consent and human subject protection, which are detailed below.

Likewise, the Monster Study (1939) was an unethical speech experiment conducted by psychologist Wendell Johnson and his graduate student Mary Tudor at the University of Iowa (Algahtani et al., 2018). The study involved orphaned children, who were divided into two groups: one group received positive speech therapy, while the other was subjected to negative reinforcement, being harshly criticised for speech imperfections. The latter group, including children who previously had no speech issues, developed speech anxiety and long-term psychological distress. Although unpublished, the study is infamous for its unethical treatment of vulnerable children and its lasting impact on ethical standards for psychological research. The ethical breaches in this study, particularly the harm inflicted on vulnerable participants without informed consent, contributed to the growing recognition of the need for formal ethical guidelines in human research. These unethical experiments did not immediately trigger a globally recognised ethical response. However, along with other notorious experiments summarised in Table 8.1, they influenced the development of ethical codes such as the Nuremberg Code (1947) and later the Declaration of Helsinki (1964), which emphasised informed consent, the minimization of harm, and the ethical treatment of research participants.

It was the atrocities of research carried out during World War II under the Nazi regime that catalysed a formal, systematic approach to research ethics, which subsequently influenced major guidelines such as the Declaration of Helsinki and the Belmont Report (World Medical Association, 2024; Nagai et al., 2022). This marked the first comprehensive international framework explicitly outlining the fundamental principles of ethical research, including voluntary informed consent and the obligation to minimise harm, particularly in the field of medical research involving human subjects and resulted in the establishment of the Nuremberg code in 1947 (Roelcke, 2004). This code laid the foundation for modern ethical standards by emphasising (a) the necessity of informed and voluntary consent from research participants, (b) the importance of avoiding unnecessary suffering and minimising

Table 8.1 Historical Examples of Reported Unethical Research Practices

Tuskegee Syphilis Study (1932) [Centre of Disease Control and Prevention, (n.d.)]	Conducted by the U.S. Public Health Services, involved withholding treatment from African American men who had syphilis, even after penicillin became a recognised cure for the disease. The participants were not informed about the nature of the study or the risks involved, and they were misled into thinking they were receiving medical treatment.
Monster Study (1939) [Practical Psychology, 2023]	This study aimed to investigate the effects of negative speech therapy on orphan children. Some children were told they had speech problems even if they did not have them. It caused psychological harm to the children involved, many of whom developed speech disorders as a result.
Nazi Human Experimentation (1943–1944) [Roelcke, 2004]	Physicians conducted painful and often deadly experiments on prisoners without their consent, using inhumane and questionable research standards.
Willowbrook Hepatitis Experiments (1950–1960)	These experiments involved intentionally infecting children at the Willowbrook State School in New York with hepatitis to study the progression of the disease. Parents of the children were often coerced into giving consent.
Stanford Prison Experiment (1971) [Stanford Libraries, n.d.)]	The experiment was aimed to study the psychological effects of perceived power. However, the study quickly spiralled out of control as participants assigned to the role of prison guards began to exhibit abusive behaviour toward those assigned as prisoners.
The Alder Hey Organs Scandal (1981–1996) [Bauchner & Vinci, 2001]	Various internal organs, including hearts and brains, had been removed during post-mortem investigations from children at Alder Hey Hospital in Liverpool without the knowledge or consent of parents. Although this work was carried out by one pathologist, the outcome highlights the institutional failings in ethical monitoring.
Vioxx Clinical Trial Controversy (1999) [Krumholz et al., 2007]	Vioxx (Rofecoxib), a pain medication manufactured by Merck & Co., involved allegations of withholding data about the drug's cardiovascular risks during clinical trials. The controversy raised questions about the transparency of pharmaceutical research and the reporting of adverse effects.
Facebook Emotional Manipulation (2014) [Boyd, 2014]	The study manipulated the news feeds of approximately 700,000 users to gauge their emotional responses. This was published in the Proceedings of the National Academy of Sciences, and it received widespread criticism for its lack of informed consent and potential psychological harm to participants.

risks to participants, and most importantly, (c) the requirement for researchers to conduct research where results can benefit society. This document has profoundly influenced subsequent ethical guidelines and regulations, including the Declaration of Helsinki and the Belmont Reports, which shape the principles that govern human research ethics today (Nagai et al., 2022; Carlson et al., 2004) (see also sources in Table 8.1). Since these initiatives, several changes, adaptations, and refinements were made to the Nuremberg code in an attempt to guide researchers in subject areas beyond medicine.

In 2013, the World Health Organisation (WHO) updated the Helsinki Declaration to produce the International Ethical Guidelines for Biomedical Research Involving Human Subjects (CIOMS, 2017), which helped institutions and organisations establish research ethics and approval processes to govern the standards of conduct for scientific researchers. These measures can only be implemented through compliance by all stakeholders. Ethical compliance ensures actions, decisions, and behaviours align with established ethical norms and values. It encompasses a range of practices and measures aimed at promoting ethical behaviour, preventing misconduct, and fostering a culture of integrity. In fact, "compliance" with ethics should be considered beyond following ethical laws and regulations (or more than just avoiding situations that lead to accusations of research misconduct) (European Commission, n.d.; Lyle et al., 2022). It is interesting to note that some forms of unethical practices are still being reported by many investigators (Kearney et al., 2024; Poutoglidou et al., 2022; Sivasubramaniam et al., 2021). Although these studies were conducted by anonymous questionnaires and focus group discussion, they highlight the fact that unethical practices are happening despite having rigorous ethical approval with risk assessments. The main reasons for unethical practices in research are noncompliance to ethical guidelines or negligence (Asplund & Åsberg, 2021; Wu et al., 2019).

The cases summarised in Table 8.1 serve as reminders of the importance of ethical oversight in research and the need to prioritise the well-being and rights of participants.

Other reasons for research misconduct include lack of understanding, inadequate training, personal or organisational pressures, and time constraints (i.e., tight deadlines), with or without financial incentives (such as linking to performance metrics or promotions) (Rehbock, 2011; Cornejo-Moreno & Gómez-Arteaga, 2012). As stated by Dyck and Allen (2013), the responsibility for ethical compliance should be owned by the researcher. However, due to external pressures such as the "publish or perish" culture, researchers are being socialised by the academy to become egocentric and, as a result, failing to abide by basic guidelines of ethical research (Quaia & Vernuccio, 2022; Paruzel-Czachura et al., 2020; Gandevia, 2018; Rawat & Meena, 2014).

In the light of these historical cases and how they influenced the development of the broadly accepted ethical principles now in place, the author was interested to explore the current situation by (a) capturing researchers' opinions on compliance with ethical guidelines required by funding agencies and ethical review boards; and (b) presenting recent cases exemplifying both ethical and questionable research conduct across different disciplines. The methodology adopted for the study is described next.

Methodology

The aim of the study was to explore and present some exemplary as well as non-exemplary research behaviours. The examples were gathered through an exploration of ethical and behavioural changes using informal dialogue with a small representative sample of those involved in STEM and non-STEM research from Europe, the Middle East, and South-East Asia.

Narrative interviews were used to obtain information on researchers' attitudes towards compliance to ethical standards and to uncover incidences of unreported ethical misconduct. Our previous experience showed that this methodology was the best way for collecting participants' own experiences and to better understand their experiences and behaviours. Using narrative interviews also helped us to let the interviewee control the direction, content and pace of the interview.

As the objective of this study was to understand the researchers' own experiences, as well as what they had observed in others with respect to adhering to ethical requirements by establishing acceptable ethical behaviours, we felt narrative research was the best approach to obtain meaningful data.

Participants were recruited through the researcher's international academic contacts by circulating details of the research via email. Additionally, the research team contacted the attendees of the 9th European Conference on Ethics and Integrity in Academia (2023-ECEIA) for recruitment of participants. Thirty-five participants were recruited from Europe, the Middle East, and South-East Asia, all of whom were carrying out or involved in research under the following categories: (a) Graduate students (GS)—those conducting PhD research (n = 5; three STEM and two non-STEM); (b) Post-doctoral researchers (PDR)—Junior scientist conducting research within 5 years of PhD (n = 5; three STEM and two non-STEM), including those securing permanent positions within 6 years of post-graduation; and (c) Research Scientists (RS)—Academics with more than 10 years of research experience as principal investigators (n = 15; six STEM and nine non-STEM). Interviews were carried out under the University of Derby ethical approval (ETH2021-4418). Interviews were conducted via Microsoft Teams without video to maintain anonymity. The responses were automatically transcribed using the inbuilt Microsoft tool. The topics covered in the interview are summarised in Table 8.2, with representative starter questions. The Microsoft Teams generated and transcribed interview responses were reviewed for typographical accuracy, and interesting and significant features in relation to our research questions were identified. Due to the limited sample size, manual coding was carried out, and these codes were collated under the following themes: (a) self-motivation for establishing ethical behaviours, (b) the role of peers and supervisors in enhancing ethical practices, (c) role of institutional policies, and (d) reported or experienced QRPs.

Table 8.2 Topics Discussed in the Focus Group Interviews with Example Open Questions

Topics covered	Objective(s) behind using the topics in the discussions
Self-motivation for adhering to ethical principles.	To understand the factors that "influence" researchers to abide by (or disregard) ethics in research.
	Example interview question: "How do you ensure that your actions consistently reflect your commitment to ethical principles?"
The role of a supervisor	To understand the participants' perceptions on the "influence" of their supervisor in maintaining research ethics.
	Example interview question: "Can you describe how your supervisor has influenced your approach to ethical decision-making?"
The role of institutional policies	To gauge how institutional policies (and disciplinary procedures) influence ethical behaviour amongst researchers.
	Example interview question: "How effective do you think your institutional efforts are in promoting ethical behaviour?"
Peer influences	To investigate the effects of laboratory environments, peers, and their ethical behaviours on individual abidance.
	Example interview question: "Can you discuss an experience where observing either ethical or unethical behaviour by others in your lab impacted your own ethical practices?"
The way the projects are being funded	To understand the differences in abiding ethical principles between internal (institutional) and external (including charities) funding bodies.
	Interview starter question: "What challenges have you faced in maintaining ethical standards when working with internal funding compared to external funding bodies?"
Reported and/or noticed misbehaviours	To elucidate the participants' own experience in noticing good/mal-practice and how it may have influenced their research culture.
	Interview starter question: "Can you describe a specific situation where you observed a particularly effective or ineffective practice in your work environment?"

Results and Discussion

This study aimed to (a) gather and analyse the views, opinions, and attitudes of researchers regarding ethical requirements and (b) explore researchers' first-hand experiences (both positive and negative), related to complying with ethical standards by establishing acceptable ethical behaviours in their research. The initial intention was to understand whether there are any differences between STEM and non-STEM researchers in their experiences of adhering to ethical principles by establishing acceptable ethical behaviours. This comparison was not possible due to the limited sample size. Therefore, the data is presented as a narrative illustration of the participants' own experiences under the themes outlined in Table 8.2. We subsequently compared this thematic analysis with previous findings to determine if there have been any changes in the reported trends.

Self-Motivation for Compliance with Ethical Standards

Work ethic is often attributed to hard work and industriousness (Grabowski et al., 2021). Many researchers feel "obliged" to carry out research in a manner that is expected from them–a desirable moral quality (Rensik, 2020). Also, self-motivation serves as a powerful catalyst for ethical behaviour in research. However, it is often linked to the moral duty to fulfil the expectations of the researcher's supervisor, the institution, and/or the funding body (Khosa et al., 2023). When researchers are intrinsically motivated to abide by ethical standards, they are more likely to uphold these standards even in the absence of external supervision or incentives. We examined the participants' perceptions of self-motivation for ethical compliance in research. All 35 participants agreed that self-motivation plays a vital role in ensuring adherence to ethical rules in research. The general consensus was that a commitment to ethical standards could be fulfilled by emphasising the importance of ethical conduct and aligning individual actions with intrinsic values. However, some participants ($n = 6$) pointed out that their self-motivation for ethical research is often challenged by the behaviours of their peers. One PhD student revealed:

> I always abide by the laboratory rules, but some others do not care about ethics [...]It is demoralising to me when those who produce results mostly by non-abiding the basic ethics were praised.

Another post-doctoral fellow commented:

"In research, all that matters is the data or results. My supervisor does not even care about how that data is produced". "So why should I bother about self-motivation?"

Most participants ($n = 34$), including research supervisors (RS), agreed that cultivating self-motivation to abide by ethical rules requires a multifaceted approach involving external influences.

The Role of Peers and Research Supervisors

Peer influence can significantly impact ethical behaviours in research (Malesky et al. 2021). Understanding these influences can help researchers develop strategies to mitigate unethical behaviours and promote a culture of integrity. The participants of our study have shared examples of how peer influences have positively and negatively impacted their behaviours. Sixteen participants reported working under supervisors who were good role models. One PDR explained how they became a role model for their junior peers, mainly because they never practised any QRPs. According to this participant, *"this has produced a positive working environment,"* in which everyone considered ethics as a part of their life.

One GS stated:

"The post-doc in my laboratory manages everything. She checks the data produced by us and makes sure the methods used are appropriate". "This has reassured me that the data produced by me is robust".

On the other hand, some PDRs (three out of five) reported witnessing their peers regularly cutting corners in experiments to meet tight deadlines. One of them stated:

> In a research lab, it's common practice to manipulate data to achieve statistically significant results. New members quickly learn that producing desirable results is more important than accuracy.

These PDRs have seen senior researchers frequently altering data without repercussions, leading them to believe this behaviour is acceptable and even necessary for career advancement. Further examples are explained under the theme of reported misbehaviours or QRPs (see below).

Pizzolato and Dierickx (2022) suggest that supervisors play an important role in modelling and teaching students to learn and practice research integrity. Research students see their supervisors as their role models, therefore, ethical and unethical behaviours demonstrated by supervisors are learned and internalised by their trainees. We wanted to examine this relationship further and explore the participants' opinions on the role of the supervisor in ethical training.

All thirty-five participants highlighted the importance of training students on adhering to ethical principles in their day-to-day activities. Five senior researchers (SR) in STEM fields felt that due to increasing trainee numbers, it is not always possible to guide students on ethical behaviour, which is an essential part of their research training.

One PDR respondent stated:

> In fact, the fundamental part of any studentship in higher education (HE) is the project component, and the supervisor has a pivotal role in instilling the best possible ethical standards to the next generation of scholars, irrespective of the capabilities (with respect to the field).

In this disclosure, it seems that the participant is suggesting that the supervisor serves as a guide in two ways: (1) by acting as a subject specialist and an educator who provides constructive feedback to improve the student's research skills and (b) by acting as a mentor to help guide learners in the development and practice of ethical behaviours as well as ensuring their understanding of the importance of ethical practice, as well as the consequences of non-compliance.

Another participant (RS) talked about the supervisor being a role model for ethical compliance:

> When students of any level see supervisors not following established guidelines concerning their practice, they will inevitably emulate this behaviour, passing on these poor practices to the next generation of scholars.

Training, particularly in the context of STEM disciplines, is critical for both the rigour of the work conducted and the safety of the immediate user and those who share that space.

One PDR stated:

> "Misappropriation or stealing of consumables from other researchers, imposing themselves on junior researchers, or not following legal or local laws can lead to harm as well as invalidation of any data that is generated as part of a study". "These issues create conflict in a small group, making it challenging to produce the high-quality research needed to drive a field forward."

It is worth noting that there is a close and hierarchical relationship between a trainee and their supervisor and possibly a second supervisor (Tinsley, 2022). Therefore, conflicts between a trainee and their supervisory teams are difficult to navigate. The student not only needs to be comfortable in challenging decisions by the supervisor but also willing to accept rational decisions by the supervisor/supervisory team (Gunnarsson et al., 2013).

An RS stated:

> Expectations on independence vary depending on the level of study, however, the key message to be communicated to the student is that the supervisor makes decisions based on the wealth of their experience.

During the conversation, this participant (RS) recalled an incident where a student disagreed with the method proposed by their supervisor. This resulted in the student attempting to convince other members of the research team to consider another methodology, which was not appropriate for that particular investigation. This situation created multiple conflicts amongst (a) the student and supervisor, (b) the supervisor and technical staff, (c) the student and other members of the team, and (d) the student and technical staff. These conflicts created a general sense of mistrust and tension in the environment. The resolution of this conflict required the supervisor to discipline the student, which had damaging impacts on both the student's work and the work of others.

The Role of the Institution

Some participants (mainly the GSs and PDRs) discussed the importance of institutional measures, guidelines, and training, as well as access to a person that could provide urgent ethical advice. Many participants (19 out of 35) shared examples of what they believed to be effective supports at their own institutions. This information is summarised in Table 8.3.

Table 8.3 Identified Institutional Measures to Enhance Supervisor-Researcher Relationships

Institutional Measure	How it Improved Supervisor-Researcher Relationships
Clear Procedures	Clear procedures with flow charts to get help during induction.
Independent Internal Verifier	Another named academic in the supervisory team who can be contacted during a conflict.
Institutional Ombudsman	Institutional research officer who oversees and resolves any supervisor-researcher conflicts.
Student Workshop	PhD workshop titled "Managing your supervisor," providing training to establish working relationships with supervisors.
Joint Workshop	Monthly small group discussion forum in which supervisor-student pairs attend and reflect on their relationship (in the event of persistent conflict, this can be made compulsory).
Compulsory Laboratory Meetings	Similar to joint workshops; however, the autonomy is given to the individual laboratory researchers and their supervisors to conduct weekly laboratory meetings.
Recognition and Rewards	Recognition programs that reward effective supervision and mentoring, such as awards or public acknowledgement. Incentives for supervisors to engage in professional development and to adopt best practices in supervision.
Conflict Resolution Resources	Accessible conflict resolution services, such as ombudspersons or mediation services, to address and resolve disputes between supervisors and researchers. Clear, confidential, and non-punitive processes for handling grievances.
Workload Management	Ensure supervisors have manageable workloads to allow adequate time for mentoring and supporting their researchers. Monitor and regulate the number of researchers assigned to each supervisor to maintain effective supervision.

These examples are based on interview responses given by GS, PDRs and RS from different institutions

Reported Misconducts or QRPs

The second part of the interviews focussed on inappropriate behaviours and misconduct. Researchers are facing extreme pressures to conduct investigations with reproducible and robust data (House of Commons Science, Innovation and Technology Committee, 2023). Authors from all over the world are still reporting evidence for unethical practices, research, and publication misconduct (Fink et al., 2023; Bouter, 2023; Rodrigues et al., 2023; Gøtzsche, 2022). In 2021, one of us reported on unethical practices in STEM related research (Sivasubramaniam et al., 2021). In the current study, we wanted to explore this further.

We queried our participants about whether they had conducted or witnessed QRPs. QRPs investigated in the current study included subtle data manipulation such as P-hacking (manipulation of data analysis to achieve statistically significant results) (Head et al., 2015), ignoring ethical standards, selective reporting of data, hypothesising after the results are known (HARKing), duplicate publications, Salami slicing (dividing a single significant research study into multiple smaller papers to increase the number of publications), or dividing one significant piece of research into multiple smaller publications to increase the number of publications, often without appropriate disclosure, and improper authorship, and failure to report conflicts of interest, etc. Unsurprisingly, all thirty-five participants only reported on witnessed behaviours, none reported on their own QRPs. Data manipulation and/or fabrication was reported as being witnessed by twenty-seven of the thirty-five participants. STEM participants narrated examples of unreported experimental failures (positivity bias). Data manipulation or falsification of data is a significant issue and has resulted in several high-profile instances of retractions of research over the years (Lievore et al., 2021). As previously stated, students should be able to report to their supervisor when something is not working as originally hypothesised. Arguably, this is the basis of scholarly activity, as it is necessary to "follow the data" and interpret data that is observed rather than fitting the data to a pre-existing narrative. A STEM PDR in this study recalled an incident in which students were working on a project as part of a broader team. One student within the team confided with the other that they were concerned their data was not "following" their hypothesis, and they were worried that the work they had conducted was not adequate. They then proceeded to repeat the experiment with some methodological adjustments and subsequently presented alternative results. This PDR further added:

> In my opinion, it was important to address this concern before it got out of control, as this is damaging from both an ethics and scientific perspective [...]. Scientifically, it will perpetuate a position which is not progressing the field forward, leading to misallocation of time and resources. Ethically, this level of deception results in a loss of trust between student and supervisor and may impact on others involved in the work and impact on the expansion of the work moving forward.

Additionally, one Research Scientist (RSci) described an incident in which an undergraduate student failed to follow aseptic techniques, which resulted in contaminated cell culture plates. Despite noticing clear signs of contamination, they did not inform the supervisor and instead continued to complete the experiment with contaminated cells. The data obtained was then analysed and presented to the supervisor. Fortunately, the supervisor was able to identify the contamination from skewed values. The RS further added that simple incidences/research malpractices like this may cause severe consequences.

As previously described, a few participants (two GSs, five PDRs and nine RSci's) reported on the positive and negative influences of their peers in abiding by ethical guidelines or behaving with moral conscience. In contrast, one PhD student (GS) stated:

> I was annoyed to see how one of my peers manipulated the data but still got published [...]. This has negatively impacted on my behaviour, too!

Similar stories have been narrated by other participants. These are summarised in Table 8.4, which categorises responses about peer influences as either positive (i.e., acceptable practices) or negative (QRPs). PDRs who are now working as early career academics (ECA) have identified issues of plagiarism and misuse of Artificial Intelligence (AI) as a main concern. They are worried that the students are relying

Table 8.4 Summary of Identified Positive and Negative Peer Influences on Research Conduct

Positive Peer Influences	Negative Peer Influences
Collaboration and Knowledge Sharing Senior PhD students, post-doctoral fellows, and academics providing guidance to peers by explaining the reasons behind adhering to ethical principles.	Committing Research Misconduct Peers purposefully ignoring health and safety standards and practices, manipulating data, or falsifying reports.
Peer Review and Feedback Peers acting as a model for developing skills, reviewing manuscripts, and providing constructive feedback to improve writing ethics.	Peer Pressure for Positive Results Collaborative research peers demanding publishable data, urging to ignore basic ethical behavioural expectations, or ignoring approval conditions.
Provide Informal Ethical Guidance. Peers acting as the first point of contact to discuss the ethical implications of any research procedure, making research questionnaires, or providing authorship advice.	Research Misconduct Conducting unethical procedures, such as data fabrication, plagiarism, or undisclosed or unauthorised use of AI.
Maintaining Professional Standards Peers enacting professional behaviours, including courtesy, helpfulness, and consistently, abiding by the ethical expectations of the institution and/or external funding body.	Unprofessional Behaviours Stealing data, non-disclosures, conducting interviews without proper informed consent, or expecting authorship (without major contributions).

on ChatGPT and similar AI tools in constructing project reports and thesis materials in both undergraduate (UG) and postgraduate (PG) studies.

Authorship Misconduct

Authorship disputes and misconduct can be common, where collaborative projects involve multiple contributors, including PhD students, post-doctoral fellows, and established researchers/scholars. As summarised by Sivasubramaniam et al. (2021), unethical practices related to authorship may include ghost writing, honorary authorship, or the omission of deserving contributors, all of which undermine academic integrity and fairness. The responses collected from participants of this study align with previous findings (Khezr & Mohan, 2022; Rensik, 2022; Sivasubramaniam et al., 2021). A PDR in this investigation expressed concerns over inadequate acknowledgement of contributions, particularly among research team members.

They reported:

> Since this is my first academic position and the fact that I am still in the probationary contract period, I was denied of authorship of a paper to which I contributed–50% of the data presented were obtained by me!

This type of unfair treatment highlights that the issues are not only at the supervisory level but are also encouraged by the institutions. Seven participants reported on institutional malpractices relating to inclusivity in research excellence.

These responses were mainly from UK participants relating to Research Excellence Framework (REF) exercises, in which universities are evaluated based on the quality of research. It should be noted that scholars from other parts of the world narrated similar stories.

One UK-based RS pointed out a common issue regarding institutional malpractice in including researchers in the REF submission:

> Since my research outputs have medium impact, I have not been included in the REF submission, although I maximally contributed for several research outputs.

Another RSci pointed out:

> The REF exercise itself is flawed." "Universities are expected to compile a list of research-active staff based on a minimum 20% contribution to research activities, but I have witnessed some institutions deliberately lowering this percentage for academics with less impactful outputs.

Although this is an accepted practice, many academics have criticised the REF exercise as not fit for purpose as it undermines collegiality and innovation (Sayer, 2014; Gadd, 2022). Newspapers have reported the stories of unsatisfied academics who questioned the competence of REF panels in decision making, or others who questioned the procedures used by the individual universities to select academics/researchers for REF submissions. The participant responses from our study support this finding of inequality. It is important to highlight that the pre-requisite for REF submission requirement of "academics with 20% or more workload" as research related activity, is not going to change in the near future.

Concluding Remarks and Suggestions for Enforcing Compliance

The main objective of this study was to highlight those research behaviours and activities that are unethical and lead to research misconduct. The fact that some of the unethical behaviours previously reported in the literature are still being described as a common occurrence highlights the pressing need for a cultural shift amongst researchers. Factors such as high expectations ("publish or perish" culture and/or individual ambitions), lack of training, and/or nonadherence to institutional policies often result in research misconduct. While the sample size in this study was a limitation, the data presented are consistent with issues highlighted by previous authors, supporting the need for change. We propose the following as actions to support this change and minimise unethical practices in research.

- **Clear Institutional Policies**: Establishing clear institutional guidelines that articulate the ethical standards and expectations, and highlighting the consequences for non-compliance is essential. These guidelines should be clearly communicated to establish a culture of integrity and accountability. Policies should offer resources to researchers on ethical issues, such as obtaining informed consent, handling data responsibly, avoiding conflicts of interest, and following regulatory requirements. This type of support will help researchers to navigate ethical dilemmas and make informed decisions.

- **Continuous Education/Training on Ethics**: Ethics training helps researchers recognise common ethical issues and challenges that may arise during the research process. This includes issues related to authorship, data management, conflicts of interest, and publication practices. By raising awareness, researchers can proactively address and minimise the risk of misconduct. Training should be focused on situational ethical decision making in complex situations, with emphasis on the importance of transparency, rigor, and accountability in research conduct.
- **Regular Research Group Meetings to Discuss Ethical Issues**: Meetings provide an opportunity to not just reflect on and report ethical dilemmas faced by individual researchers, but meetings can also provide the space to gain advice from peers and supervisors. Through this practice, a culture of integrity can be created. These meetings will enforce a shared commitment to upholding ethical standards among researchers and deter misconduct.
- **Continuous Monitoring and Annual Reporting:** Continuous monitoring and annual reporting are essential components of institutional efforts to deter research misconduct and promote research integrity. This should involve a systematic review of research activities, data, and outputs to identify any irregularities or suspicious patterns that may indicate potential misconduct. By monitoring research processes in real-time, institutions can detect misconduct at an early stage, preventing its escalation and minimising its impact.

By prioritising continuous monitoring, robust oversight mechanisms, and ongoing education/training initiatives, institutions can foster a culture of ethical conduct and accountability, thereby safeguarding the welfare of the researchers, the research participants as well as the integrity of research results. Moving forward, sustained efforts to reinforce compliance with ethical principles are imperative to mitigate the risk of misconduct and uphold the highest standards of research integrity.

References

Algahtani, H., Bajunaid, M., & Shirah, B. (2018). Unethical human research in the field of neuroscience: A historical review. *Neurological Sciences, 39*, 829–834. https://doi.org/10.1007/s10072-018-3245-1

Altman, L. K. (2006). For science gatekeepers, a credibility gap. *The New York Times*. Retrieved from http://www.nytimes.com/2006/05/02/health/02docs.html?pagewanted=all. Accessed 26 July 2019.

Asplund, K., & Åsberg, K. H. (2021). Reporting ethical approval in health and social science articles: An audit of adherence to GDPR and national legislation. *BMC Medical Ethics, 22*, 92. https://doi.org/10.1186/s12910-021-00664-w

Baker, M. (2016). 1,500 scientists lift the lid on reproducibility. *Nature, 533*, 452–454. https://doi.org/10.1038/533452a

Bauchner, H., & Vinci, R. (2001). What have we learnt from the Alder Hey affair? *BMJ, 322*(7282), 309–310. https://doi.org/10.1136/bmj.322.7282.309

Bouter, L. (2023). Why research integrity matters and how it can be improved. *Accountability in Research, 31*(8), 1277–1286. https://doi.org/10.1080/08989621.2023.2189010

Boyd, D. (2014). What does the Facebook experiment teach us? Growing anxiety about data manipulation. *Medium.* https://medium.com/message/what-does-the-facebook-experiment-teach-us-c858c08e287f.

Brandt, A. M. (1978). Racism and research: The case of the Tuskegee syphilis study. *The Hastings Centre Report, 8*(6), 21–29. https://doi.org/10.2307/3561468

Carlson, R. V., Boyd, K. M., & Webb, D. J. (2004). The revision of the declaration of Helsinki: Past, present and future. *British Journal of Clinical Pharmacology, 57*(6), 695–713. https://doi.org/10.1111/j.1365-2125.2004.02103.x

Centre of Disease Control and Prevention. (n.d.). *The untreated syphilis study at Tuskegee timeline*. CDC. https://www.cdc.gov/tuskegee/timeline.htm

Cornejo-Moreno, B. A., & Gómez-Arteaga, G. M. (2012). Violation of ethical principles in clinical research. Influences and possible solutions for Latin America. *BMC Medical Ethics, 13*(35). https://doi.org/10.1186/1472-6939-13-35

Council for International Organizations of Medical Sciences (CIOMS). (2017). *International ethical guidelines for health-related research involving humans*. https://cioms.ch/wp-content/uploads/2017/01/WEB-CIOMS-EthicalGuidelines.pdf

Dyck, M., & Allen, G. (2013). Is mandatory research ethics reviewing ethical? *Journal of Medical Ethics, 39*, 517–520. https://doi.org/10.1136/medethics-2011-100274

European Commission. (n.d.). *Ethics*. H2020 Online Manual. https://ec.europa.eu/research/participants/docs/h2020-funding-guide/cross-cutting-issues/ethics_en.htm

Fanelli, D. (2009). How many scientists fabricate and falsify research? A systematic review and meta-analysis of survey data. *PLoS One, 4*(5), e5738. https://doi.org/10.1371/journal.pone.0005738

Fink, M., Gartner, J., Harms, R., & Hatak, I. (2023). Ethical orientation and research misconduct among business researchers under the condition of autonomy and competition. *Journal of Business Ethics, 183*, 619–636. https://doi.org/10.1007/s10551-022-05043-y

Gadd, E. (2022, May 26). Tales from REF Central: Reflections from REF results week. *HEPI.* https://www.hepi.ac.uk/2022/05/26/tales-from-ref-central-reflections-from-ref-results-week/

Gandevia, S. (2018). Publication pressure and scientific misconduct: Why we need more open governance. *Spinal Cord, 56*, 821–822. https://doi.org/10.1038/s41393-018-0193-9

Gøtzsche, P. C. (2022). Citation bias: Questionable research practice or scientific misconduct? *Journal of the Royal Society of Medicine, 115*(1), 31–35. https://doi.org/10.1177/01410768221075881

Grabowski, D., Chudzicka-Czupała, A., & Stapor, K. (2021). Relationships between work ethic and motivation to work from the point of view of the self-determination theory. *PLoS One, 16*(7), e0253145. https://doi.org/10.1371/journal.pone.0253145

Gunnarsson, R., Jonasson, G., & Billhult, A. (2013). The experience of disagreement between students and supervisors in PhD education: A qualitative study. *BMC Medical Education, 13*, 134. https://doi.org/10.1186/1472-6920-13-134

Head, M. L., Holman, L., Lanfear, R., Kahn, A. T., & Jennions, M. D. (2015). The extent and consequences of p-hacking in science. *PLoS Biology, 13*(3), e1002106. https://doi.org/10.1371/journal.pbio.1002106

House of Commons Science, Innovation and Technology Committee. (2023). *Reproducibility and research integrity - Sixth report of session 2022–23.* https://committees.parliament.uk/publications/39343/documents/194466/default/

Ioannidis, J. P. A. (2005). Why most published research findings are false. *PLoS Medicine, 2*(8), e124. https://doi.org/10.1371/journal.pmed.0020124

John, L. K., Loewenstein, G., & Prelec, D. (2012). Measuring the prevalence of questionable research practices with incentives for truth. *Psychological Science, 23*(5), 524–532. https://doi.org/10.1177/095679761143095

Kearney, M., Downing, M., & Gignac, E. A. (2024). Research integrity and academic medicine: The pressure to publish and research misconduct. *Journal of Osteopathic Medicine, 124*(5), 187–194. https://doi.org/10.1515/jom-2023-0211
Khezr, P., & Mohan, V. (2022). The vexing but persistent problem of authorship misconduct in research. *Research Policy, 51*(3), 104466. https://doi.org/10.1016/j.respol.2021.104466
Khosa, A., Wilkin, C., & Burch, S. (2023). PhD students' relatedness, motivation, and well-being with multiple supervisors. *Accounting Education, 33*(2), 131–163. https://doi.org/10.1080/09639284.2023.2179889
Krumholz, H. M., Hines, H. H., Ross, J. S., Presler, A. H., & Egilman, D. S. (2007). What have we learnt from Vioxx? *BMJ, 334*, 120–123. https://doi.org/10.1136/bmj.39024.487720.68
Lievore, C., Rubbo, P., Biynkievycz dos Santos, C., Picinin, C. T., & Pilatti, L. (2021). Research ethics: A profile of retractions from world class universities. *Scientometrics, 126*, 6871–6889. https://doi.org/10.1007/s11192-021-03987-y
Lyle, K., Weller, S., Samuel, G., et al. (2022). Beyond regulatory approaches to ethics: Making space for ethical preparedness in healthcare research. *Journal of Medical Ethics, 49*, 352–356. https://doi.org/10.1136/medethics-2021-108102
Malesky, A., Grist, C., Poovey, K., & Dennis, N. (2021). The effects of peer influence, honor codes, and personality traits on cheating behavior in a university setting. *Ethics & Behavior, 32*(1), 12–21. https://doi.org/10.1080/10508422.2020.1869006
Mandal, J., Acharya, S., & Parija, S. C. (2011). Ethics in human research. *Tropical Parasitol, 1*(1), 2–3. https://doi.org/10.4103/2229-5070.72105
Munafò, M. R., Nosek, B. A., Bishop, D. V. M., Button, K. S., Chambers, C. D., du Sert, N. P., & Ioannidis, J. P. A. (2017). A manifesto for reproducible science. *Nature Human Behaviour, 1*(1), 0021. https://doi.org/10.1038/s41562-016-0021
Nagai, H., Nakazawa, E., & Akabayashi, A. (2022). The creation of the Belmont Report and its effect on ethical principles: A historical study. *Monash Bioethics Review, 40*, 157–170. https://doi.org/10.1007/s40592-022-00165-5
National Academies of Sciences, Engineering, and Medicine. (2017). F*ostering integrity in research.* Committee on Science, Engineering, Medicine, and Public Policy; Committee on Responsible Science National Academies Press (US). http://dx.crossref.org/10.17226/21896
Nelson, N. C., Ichikawa, K., Chung, J., & Malik, M. M. (2021). Mapping the discursive dimensions of the reproducibility crisis: A mixed methods analysis. *PLoS One, 16*(7), e0254090. https://doi.org/10.1371/journal.pone.0254090
Parchomovsky, G. (2000). Publish or perish. *Michigan Law Review, 98*(4), 926–952. https://doi.org/10.2307/1290335
Paruzel-Czachura, M., Paruzel-Czachura, M., Baran, L., & Spendel, Z. (2020). Publish or be ethical? Publishing pressure and scientific misconduct in research. *Sage Journals, 17*(3). https://doi.org/10.1177/174701612098056
Pizzolato, D., & Dierickx, K. (2022). Research integrity supervision practices and institutional support: A qualitative study. *Journal of Academic Ethics, 21*, 427–448. https://doi.org/10.1007/s10805-022-09468-y
Poutoglidou, F., Stavrakas, M., Tsetsos, N., Poutoglidis, A., Tsentemeidou, A., Fyrmpas, G., & Karkos, P. D. (2022). Fraud and deceit in medical research: Insights and current perspectives. *Voices in Bioethics, 8*. https://doi.org/10.52214/vib.v8i.8940
Quaia, E., & Vernuccio, F. (2022). Finding a good balance between pressure to publish and scientific integrity and how to overcome temptation of scientific misconduct. *Tomography, 8*(4), 1851–1853. https://doi.org/10.3390/tomography8040155
Rawat, S., & Meena, S. (2014). Publish or perish: Where are we heading? *Journal of Research in Medical Sciences, 19*(2), 87–89. PMID: 24778659; PMCID: PMC3999612. https://api.semanticscholar.org/CorpusID:4857778
Rehbock, T. (2011). Limits of autonomy in biomedical ethics?: Conceptual clarifications. *Cambridge Quarterly of Healthcare Ethics, 20*(4), 524–532. https://doi.org/10.1017/S0963180111000260

Resnik, D. B. (2020, December 3). What is ethics in research & why is it important? *National Institute of Environmental Health Sciences (NIH) Article.* https://www.niehs.nih.gov/research/resources/bioethics/whatis

Rensik, D. (2022). Should authorship on scientific publications be treated as a right? *Journal of Medical Ethics, 49, 776–778.* https://doi.org/10.1136/jme-2022-108874

Rodrigues, F., Gupta, P., Khan, A. P., Chatterjee, T., Sandhu, N. K., & Gupta, L. (2023). The cultural context of plagiarism and research misconduct in the Asian region. *Journal of Korean Medical Science, 38*(12), e88. https://doi.org/10.3346/jkms.2023.38.e88

Roelcke, V. (2004). Nazi medicine and research on human beings. *Lancet, 364*(Suppl 1), s6–s7. https://doi.org/10.1016/S0140-6736(04)17619-8

Sayer, D. (2014, December 15). Five reasons why the REF is not fit for purpose. *The Guardian.* https://www.theguardian.com/higher-education-network/2014/dec/15/research-excellence-framework-five-reasons-not-fit-for-purpose

Sivasubramaniam, S., Dlabolová, D. H., Kralikova, V., et al. (2021). Assisting you to advance with ethics in research: An introduction to ethical governance and application procedures. *International Journal for Educational Integrity, 17,* 14. https://doi.org/10.1007/s40979-021-00078-6

Sivasubramaniam, S. D., Cosentino, M., Ribeiro, L., & Marino, F. (2021). Unethical practices within medical research and publication—An exploratory study. *International Journal for Educational Integrity, 17*(7). https://doi.org/10.1007/s40979-021-00072-y

Snezhko, S., & Coskun, A. (2019). Liability or ethics? The real value of compliance. In U. Akkucuk (Ed.), *The circular economy and its implications on sustainability and the green supply chain* (pp. 213–230). https://doi.org/10.4018/978-1-5225-8109-3.ch011

Stanford Libraries. (n.d.). Stanford prison experiment August 15–21, 1971. Stanford University. https://exhibits.stanford.edu/spe

Tinsley, E. A. (2022). *Nondisclosure of supervisee disagreement to supervisor feedback: A qualitative study.* Dissertations (1934 -) 2572. https://epublications.marquette.edu/dissertations_mu/2572

Titus, S., Wells, J., & Rhoades, L. (2008). Repairing research integrity. *Nature, 453*, 980–982. https://doi.org/10.1038/45390a

World Medical Association. (2024). Declaration of Helsinki – Ethical principles for medical research involving human participants. https://www.wma.net/policies-post/wma-declaration-of-helsinki/

Wu, Y., Howarth, M., Zhou, C., et al. (2019). Reporting of ethical approval and informed consent in clinical research published in leading nursing journals: A retrospective observational study. *BMC Medical Ethics, 20*(1), 94. https://doi.org/10.1186/s12910-019-0431-5

Chapter 9
The Journey to Explicit Ethics Education for Czech Computer Science Teaching Assistants

Tomáš Foltýnek and Martin Ukrop

Abstract Student teaching assistants have two roles simultaneously—students and teachers—which leads to inherent conflict of interest. This chapter shares the experience of the Faculty of Informatics, Masaryk University in Czechia, where ethics education is not obligatory for teaching assistants. Over multiple semesters, we conducted surveys on attitudes toward ethical issues in teaching practice. Based on the survey results, we developed the teaching toolkit on ethical dilemmas, including the framework for ethical decision-making. We incorporated it into the teachers' education for computer science teaching assistants. We also developed a Code of Ethics for Tutors, which became an official faculty document. Altogether, these results help to raise awareness of ethical issues that student teaching assistants face and positively contribute to the culture of academic integrity. We discovered that computer science student teaching assistants are most hesitant about their relationships with students and conflicts of interest arising from their dual roles. Our findings highlight the importance of incorporating ethics into teacher education and increasing awareness of these issues across faculty. They can serve as inspiration on how to explicitly include ethics in the teachers' education and raise awareness of ethical issues across the Faculty.

Keywords Ethics education · Teaching assistants · Code of ethics · Ethical decision-making

T. Foltýnek (✉) · M. Ukrop
Masaryk University, Brno, Czechia
e-mail: foltynek@fi.muni.cz; mukrop@mail.muni.cz

I. Glendinning et al. (eds.), *Ethics and Integrity in Education (Research)*, Ethics and Integrity in Educational Contexts 9,
https://doi.org/10.1007/978-3-031-98157-9_9

Introduction

For the academic community of the Faculty of Informatics, Masaryk University, Czechia (FI MU), academic integrity is taken seriously, not just as a policy but as a core part of our academic culture, even though, from time to time, the Faculty experienced unethical behaviour committed by a teaching assistant or even an academic staff member. Cases of unethical conduct have included leaking exam questions to the students, ridiculing students or otherwise reducing their dignity. These cases were either resolved internally by the course coordinator or passed to the faculty disciplinary committee. The academic community of the Faculty agrees that these cases are unacceptable and contradict the overall institutional culture of academic integrity. The situation is further complicated by the fact that a non-negligible number of people involved in teaching are student teaching assistants (TAs), with inherent conflicts of interest due to their dual role in the system.

However, until 2022, there was no teacher training on ethical issues at the Faculty. For student TAs, a recommended course (the *Teaching Lab*) was available to introduce them to the basics of teaching, but it did not include any topics related to teaching ethics. The academic staff are also not required to attend or pass any training on this topic. The only exception was the provision of training on appropriate research conduct, which was mandatory for all PhD students. However, this training included mainly the principles of research integrity and did not touch on ethical issues in teaching activities.

One initiative would not be sufficient to address the problems mentioned above and mitigate the risks. Building a culture of academic integrity requires longitudinal, multi-faceted efforts involving multiple interconnected interventions (Bretag et al., 2014). Our approach reflects this recommendation and is based on several initiatives. Our main goal was to explicitly include ethics in the teaching practice at the Faculty. We decided to (1) draft the ethical guidelines for (student) TAs and (2) develop a teaching toolkit on ethical dilemmas, which we would include in the Teaching Lab course. To be able to draft the guidelines and design the session, we first had to answer the following (research) questions:

1. What ethical issues do the TAs struggle with most (i.e., cannot judge these issues or judge them differently)?
2. What ethical issues relevant to TAs need to be included in the Code of Ethics?

To answer these questions, we conducted a questionnaire survey among the student TAs visiting the Teaching Lab and the wider Teaching Lab community. In the following sections, we present the details of the survey and the findings most relevant to our research questions. Then, we describe the teaching toolkit on ethical dilemmas and share experiences from several session runs. The final part of the chapter presents the work on the Code of Ethics for Tutors. Thus, this chapter is an experience report of our multi-stage efforts, which involved assessing the problem via a questionnaire and then—based on the questionnaire results—solving the problem

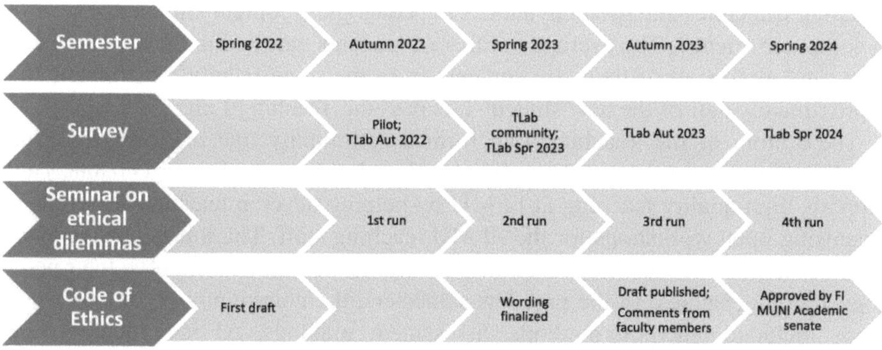

Fig. 9.1 Timeline of the activities

through the Code of Ethics and training activities tailored to the needs of computer science teaching assistants. The timeline of these activities can be seen in Fig. 9.1.

Background

This section discusses the situation around student TAs at FI MU and its national cultural context.

Student Teaching Assistants at FI MU

There are approx. 2300 students at FI MU and over 500 employees involved in teaching, one-third of which are students. On average, there are 34 new student TAs each semester. Most TAs teach one seminar group (2 h) per week, with some teaching two or three seminar groups. A higher teaching load is exceptional.

Including students as TAs in the teaching process has its advantages, such as a positive effect on student learning performance (Erdei et al., 2017; Pivkina, 2016), better approachability for students compared to professors or senior lecturers (Decker et al., 2006; Ren et al., 2019) and higher student satisfaction (Decker et al., 2006). However, there are several disadvantages, such as a potentially lower content knowledge (Riese et al., 2021) and a higher possibility of a conflict of interest (Riese & Kann, 2020).

To mitigate these risks and to ensure the high quality of teaching, FI MU offers the Teaching Lab (Ukrop, 2022; Ukrop & Švábenský, 2020). In short, Teaching Lab is an elective, semester-long course (for 3 ECTS credits) with regular weekly sessions, led by senior teaching staff, aiming to facilitate teaching reflection of novice TAs, provide teaching inspiration and build a TA community. The topics discussed range from interaction with the student group and posing good questions via basic coaching principles and group dynamics to pedagogical content knowledge specific to computer science. The average number of students enrolled in the Teaching Lab is 18, and almost all of them are students from the same faculty. This means that approximately half of the new student TAs pass the Teaching Lab.

The alumni of the Teaching Lab form a community that organises informal events, maintains communication channels (Facebook group, Discord channel) and supports high-quality teaching at FI MU by helping develop teaching materials or organising open workshops for the FI MU teaching staff. The topics change every semester to provide varied opportunities for FI MU teachers to improve their pedagogical competencies. In the past few semesters, the community-organised workshops included teaching methods, assessment methods, AI in education, and non-violent communication.

National Cultural Context

Unlike primary and secondary school teachers, higher education teachers in Czechia are not required to have any pedagogical education. Even though this is seen as an issue (Vlková, 2019), there are currently no attempts to introduce obligatory pedagogical training for higher education teachers. Therefore, systemic education of TAs and regular workshops for teaching staff at FI MU are rather exceptions in the Czech academic environment. Although student TA training sessions are standard in Western Europe and the USA (Ukrop, 2022), their context and implementation can differ a lot even within Europe (Riese et al., 2021).

Czechia belongs to the Eastern European (EE) cultural context (Foltýnek & Dlabolová, 2020). Using Hofstede's (2001) cultural dimensions, there are notable differences between EE and English-speaking countries. EE countries show higher power distance, reflecting a more hierarchical society, which is one of the causes of higher corruption (Osipian, 2017). Also, the perception of collectivism is higher in EE countries as opposed to more individualistic English-speaking countries. Students are, therefore, more inclined to share learning materials without considering this as cheating (Stöckelová & Virtová, 2015). Compared to the UK, Czech students tend to have less knowledge and more negative perceptions of plagiarism policy and procedures. They are also less willing to receive training on plagiarism issues (Mahmud et al., 2018). In the past few decades, there have been numerous initiatives aiming to improve the culture of academic integrity, namely in the area of plagiarism prevention, promotion of scientific integrity, and addressing the misconduct of academic staff (Foltýnek et al., 2024). These efforts mitigate the differences between Czechia and Western Europe, and Czech academic communities are becoming more sensitive to ethical issues and social safety. As a result, Czech universities tend to address these issues more than before.

As of 2024, Czechia does not have a national code of ethics—neither for teachers nor for researchers. The only attempt at a national-level code is the *Example Code of Ethics for Academics*, published by the Council of Higher Education Institutions in 2007. At that time, several universities adopted this code for their institution, but most of the university codes have changed since then. Most of the initiatives in the area of academic integrity are bottom-up and mainly focused on students (Foltýnek et al., 2024).

Survey on Ethical Dilemmas

To identify controversial issues worth discussing and including in the ethical guidelines, we conducted a study based on a survey of psychology graduate TAs administered at the University of Colorado in Denver (Branstetter & Handelsman, 2000). The questionnaire contained basic demographic information followed by a list of 50 scenarios. The respondents assessed each scenario on a five-point Likert scale: *"Definitely is an ethical problem," "Probably is an ethical problem," "Neutral," "Probably is not an ethical problem," "Definitely is not an ethical problem,"* with an extra option of *"I cannot judge"*. In the end, the respondents were given an opportunity to answer two open-ended questions: "*Which question(s) did you hesitate the most?*" and "*Can you think of any ethically problematic situation that you got into as a teacher?*"

Ethical issues are sensitive to verbal formulation and thus can be prone to misunderstandings, especially in non-native languages. To mitigate this risk, we provided the questionnaire in the respondents' first language. Therefore, we asked a specialist from the Department of Languages to translate the original English questionnaire into Czech. The questionnaire was piloted on five teachers of the Teaching Lab. Based on the pilot testing, we removed several items identified as irrelevant to our context and added some extra scenarios based on our faculty's previous experience. We also changed the wording of some items and adjusted the scale description to make it clearer for respondents and more suitable for our case. The complete survey instrument is available in the appendix.

After using the survey for the Teaching Lab class in autumn 2022 (14 students), we invited the wider Teaching Lab community (217 alumni of the course, including some faculty employees) to fill out the same questionnaire. We received 47 responses out of 231 (~20% response rate). Then, we used the same questionnaire in the following Teaching Lab sessions (spring 2023, 13 students; autumn 2023, 8 students; spring 2024, 9 students). In total, we received 75 responses. Almost all of the participants also took their time to answer open-ended questions, which indicates the importance of ethical issues for the respondents. As all the questions were optional, all statistics presented below are non-empty answers only.

The final sample included 20 bachelor students, 38 master students, 6 PhD students, 4 members of the academic staff and 5 faculty alumni. They were involved in teaching at the faculty on average for 4.6 semesters (std. deviation 6.2), with a median of 2 semesters and a mode of 1 semester (19 respondents). The most prevalent teaching activities were preparing/grading homework assignments (57 respondents) and tutoring a seminar alone or in pairs (51 and 52, respectively). Only a minority of respondents ran student consultation sessions (17) or directly gave lectures (9).

Just a few respondents did not indicate their ethical opinions by selecting the "*I cannot judge*" option (or, exceptionally, not answering the question). Across all 50 attitude questions, the mean count of such answers was 3.2 (std. dev. 3.0, median 3), with only two questions not being answered by more than 10 people:

- Doing as instructed by a professor, regardless of your disagreement or ethical concern: 13 participants were not able to judge
- Teaching when too distressed to be effective: 11 participants could not judge

Quantitative Results

The top five most unethical scenarios (in respective order), according to the respondents, were (Fig. 9.2):

1. Insulting, ridiculing, and so forth a student in the student's presence
2. Insulting, ridiculing, and so forth a student in the student's absence
3. Leaking confidential information about the upcoming exam to students in their seminar group
4. Teaching while under the influence of cocaine or other illegal drugs
5. Allowing a student's "likability" to influence your grading

Both the mode and median of all these five scenarios were "*Definitely is an ethical problem*".

The top five scenarios (in respective order) that the respondents did not consider as an ethical problem were:

1. Becoming sexually involved with a student only after he or she has completed your course (mode: "Definitely is not an ethical problem", median: "Definitely is not an ethical problem")
2. Encouraging students to participate in your research projects (mode: "Definitely is not an ethical problem", median: "Probably is not an ethical problem")
3. Engaging in a sexual relationship with a professor or other faculty member in your department (mode: "Definitely is not an ethical problem", median: "Probably is not an ethical problem")
4. Teaching without experience or training (mode: "Probably is not an ethical problem", median: "Probably is not an ethical problem")
5. Encouraging competition between students (mode: "Probably is not an ethical problem", median: "Probably is not an ethical problem")

The top five situations with the largest variance:

1. Using cocaine or other illegal drugs in your private (nonteaching) life (mode: "Definitely is not an ethical problem", median: "Probably is not an ethical problem")

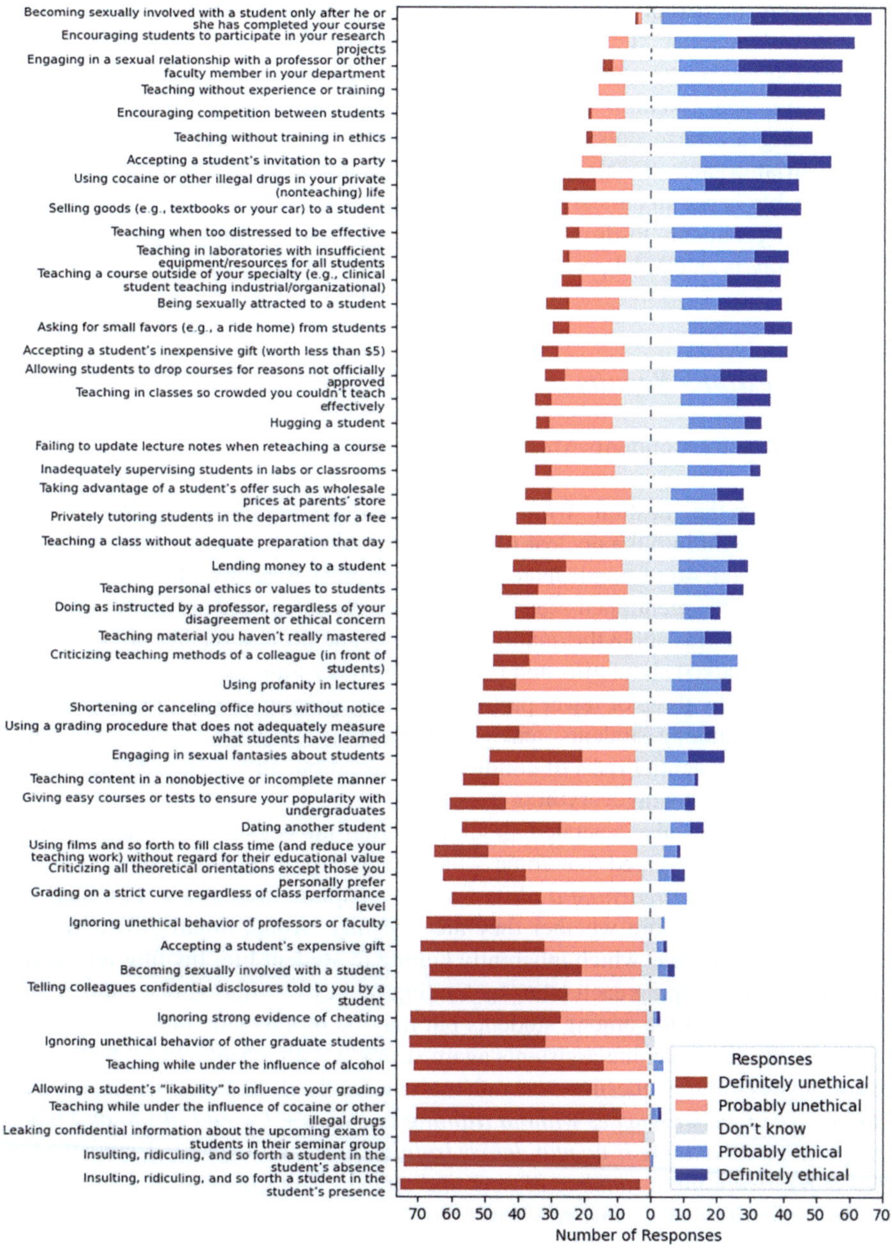

Fig. 9.2 Quantitative results of the attitude questions. Unethical towards the left, ethical towards the right. *"I cannot judge"* answers excluded

2. Engaging in sexual fantasies about students (mode: "Definitely is an ethical problem", median: "Probably is an ethical problem")
3. Being sexually attracted to a student (mode: "Definitely is not an ethical problem", median: "Neutral")
4. Teaching a course outside of your speciality (e.g., clinical student teaching industrial/organisational) (mode: "Probably is not an ethical problem", median: "Probably is not an ethical problem" /" Neutral")
5. Allowing students to drop courses for reasons not officially approved (mode: "Probably is an ethical problem", median: "Neutral")

Open Questions Results

Questions the Respondents Most Hesitated on

The answers to the question "*What question(s) did you hesitate the most?*" bring an interesting insight into the unclear topics and explain the situations with the largest variance. In total, 46 respondents took the opportunity to answer this question. Out of them, 17 (37%) mentioned scenarios about relationships and sexuality, 14 (30%) mentioned scenarios about the quality of teaching and professionality of the teacher, and 11 (24%) talked about gifts and favours. The other topics were relatively minor, i.e., mentioned by three or fewer participants. Engaging in (sexual) relationships is very common among students, and young teaching assistants are students. They understand the potential risks of conflicts of interest but still want to be part of the student community with everything it includes.

The respondents also hesitated about the borderline between ethicality and professionalism. They agreed that some scenarios like "*Teaching a course outside of your speciality*" or "*Teaching a class without adequate preparation that day*" are wrong, but weren't sure whether they are *ethically* wrong. Small gifts and other favours are other areas which inherently belong to student life. Inviting fellow students to parties, helping each other by lending money or offering discounts in parents' shops is common among students but may be problematic in teacher-student relations. Student teaching assistants again struggle with a conflict of interest. One of the answers to this question was rather about feedback on the activities we were doing: "*The teacher teaches without training in ethics + I would like to add that it is a great pity that there is no mandatory training for new TAs at FI, despite the fact that a relatively large number of new and student TAs are recruited every year, which can lead to some unwanted problems.*"

Experience with Ethically Problematic Situations

There was a great variety of answers to the question, "*Can you think of any ethically problematic situation that you got into as a teacher?*" An objective evaluation is more difficult in this case, but we can comment on the frequently mentioned topics. Relationships with students and conflicts of interest were common. The student TAs struggle with setting a clear borderline between personal and professional relationships and how friendly they should be to students. They are unsure how to solve the situation when a student asks for more points, a well-justified deadline extension, personal consultations or other advantages that might be unfair to other students. One story provided a very clear dilemma: *"I wanted to ask a student from my course on a date, but I wasn't sure if I could be objective."*

Other complicated situations include their students' personal troubles—mental breakdowns or other difficult situations. Some answers related to professional conduct, such as unpreparedness for a seminar, were mentioned several times. One respondent mentioned an issue of time dedication: *"For me, the worst dilemma is when there is one student who is completely lost and ten students who just need a little help with a minor jam in the exercise (where I help students with independent work). Then I can't decide whether I'd rather help the group of ten or devote myself to the lost student, with whom I have to spend the entire 20-30 minutes reserved for independent work in order to "catch up" with the others, and thus I won't get to the others at all."* There were also some concerns about the objectivity of grading and taking into account the attendance of classes, the number of points the students need to pass the course, or whether they repeat the course. One of the senior employees shared a very peculiar case: "*When the mother of an FI student persuaded me by e-mail to give him some advantage in his studies, and, at the same time, she teaches my children at secondary school...*".

Survey Limitations

Multiple biases may threaten the validity of our results. Firstly, we investigated teacher seniority as a potential biasing factor (as our sample was not representative of all the faculty teachers). We checked for a monotonic correlation between the number of semesters the participant has been involved in teaching at the faculty and individual attitude questions (Spearman's rank-order correlation). The correlation was significant in only three of the fifty questions:

- Ignoring strong evidence of cheating: $r_s(70) = 0.28$, $p = 0.017$;
- Teaching material you haven't really mastered: $r_s(67) = 0.28$, $p = 0.019$;
- Teaching while under the influence of cocaine or other illegal drugs: $r_s(66) = -0.29$, $p = 0.016$;

These results suggest only a weak correlation between the selected three ethical attitudes and the number of semesters the respondent has been teaching.

Secondly, the attitudes may have also been biased by having published the Code of Ethics for Tutors at the Faculty. Although this seems unlikely, as we have not done much public promotion of the Code, we checked for the bias effects by comparing the survey responses from after the publication of the Code of Ethics for Tutors draft (cohorts from semesters Autumn 2023 and Spring 2024, n = 17) with the rest of the group. According to the Kruskal-Wallis H-test, which compares the distribution of variables, only two out of fifty attitude questions have shown significant differences:

- Selling goods to a student: $\chi^2(1) = 7.20$, $p = 0.007$;
- Leaking confidential information about the upcoming exam to students in their seminar group: $\chi^2(1) = 7.54$, $p = 0.006$.

Both skewed towards being seen more as ethically problematic before the publication of the code. However, these few cases could be caused by the small sample size in one of the groups.

Thirdly, we attempted to mitigate the effects of other biases commonly present with user studies. To combat the self-selection bias during the semester survey runs, we encouraged all students to fill it out as we used its results during the session. In the community run, apart from emailing all the teachers, we intentionally approached multiple people of differing seniority and nudged them personally to make sure we'd capture a wide variety of opinions and attitudes. To mitigate the social-desirability bias, the survey was anonymous and built using a familiar tool regularly used at the university.

All in all, multiple biases were intentionally mitigated, and the others were inspected, which did not seem to skew the results significantly. Nevertheless, future replication of the study bringing more data would be beneficial in ruling out more complex biases with higher certainty.

Comparison with the Original Research

As was already mentioned, the survey instrument was reused from the original study of Branstetter and Handelsman (2000). However, any attempts to compare their results with ours are limited for the following reasons:

- The original study researched psychology TAs, whereas we focused on Computer Science TAs.
- The original study was conducted in the USA, whereas our research took place in Czechia.
- There has been an approximately 20-year gap between the studies. During this time, the ethical attitudes of TAs could change regardless of the previous issues.
- We changed some questions and the wording of the scale.

Despite these limitations, we can observe some overlaps in the results. In Branstetter and Handelman's (2000) study, the top five most unethical situations were (in respective order):

1. *Ignoring strong evidence of cheating* (8th most problematic in our data)
2. *Teaching while under the influence of cocaine or other illegal drugs* (4th)
3. *Becoming sexually involved with a student* (10th)
4. *Teaching while under the influence of alcohol* (6th)
5. *Insulting, ridiculing, and so forth a student in the student's presence* (1st)

Czech computer science teaching assistants in the 2020s—compared to their psychology counterparts from the USA 20 years ago—are more sensitive to insulting and ridiculing students (either in their presence or in their absence) and leaking confidential information to students. On the other hand, American psychology TAs condemn ignoring strong evidence of cheating and sexual involvement with a student much more than Czech computer science TAs nowadays.

The top five least problematic situations, according to Branstetter and Handelsman (2000), were (in respective order, but note that the order is reversed from the previous paragraph):

1. *Dating another graduate student* (35th least problematic in our data)
2. *Being sexually attracted to a student* (13th)
3. *Encouraging students to participate in your research projects* (2nd)
4. *Becoming sexually involved with a student only after he or she has completed your course* (1st)
5. *Teaching a course outside of your speciality (e.g., clinical student teaching industrial/organisational)* (12th)

Here, both cohorts are in agreement about encouraging students to participate in your research projects and about becoming sexually involved with a student only after they have completed your course. Dating another student was perceived as much more problematic in our study. Also, being sexually attracted to a student and teaching a course outside of your speciality is perceived as more unethical by our participants than by the participants of the original study.

The top five scenarios with the largest variance in the original study were dating an undergraduate student, using drugs in your non-teaching life, engaging in sexual fantasies about students, privately tutoring students in your department for a fee, and encouraging students to participate in your research projects. Two of these five scenarios also appear in our top five situations with the largest variance. We can see that those scenarios involving sexuality and relationships with students, as well as private tutoring and subsequent conflict of interest, are the most uncertain ones in both cohorts.

Teaching Toolkit on Ethical Dilemmas

The teaching toolkit on ethical dilemmas is a crucial intervention ensuring that all TAs who pass the Teaching Lab course receive training on ethical issues related to the teaching profession. Several studies have shown that ethical decision-making can be taught. Teaching interventions lead to a better understanding of factors that frame ethical dilemmas (Perri et al., 2009), catalyse students' feelings of personal involvement (Wotruba, 1993) and can positively influence the ethical climate in organisations (Drumwright et al., 2015). Moreover, we used the first two runs of this session to map the most pressing ethical dilemmas of young TAs and create the Code of Ethics for Tutors.

At the time of writing this chapter, we have already conducted four runs of the session, in total, for 65 students. We first present the CHEAP framework for ethical decision-making, which is a crucial part of the session, and then the session outline.

CHEAP Framework for Making Ethical Decisions

When designing the Teaching Lab session on ethical dilemmas, we wanted to provide the students with a practical tool for resolving moral dilemmas. As the target group are computer science TAs, the best format would be an "algorithm"—a deterministic process leading to a well-justified decision. The tool should allow for a quick application because the teaching-practice dilemmas usually should be solved within seconds. Of course, we knew this would not be possible, but we wanted to create something that would best suit our target group.

There are many frameworks for ethical reasoning and decision-making, ranging from the simple ones, which don't capture all necessary aspects and peculiarities of such processes, to the complex ones, which are difficult to understand, remember, and consequently apply in practice. The proposed framework combines the *Seven-step method* developed within the *Horizon 2020 project Embassy of Good Science* (The Embassy of Good Science, n.d.) and Kidder's *Checkpoints of ethical decision-making* (Kidder, 2009). It consists of five steps; the first letters constitute the acronym CHEAP, which students can easily remember:

1. **C**ONTEXT—What are the facts, and what are just conjectures? What circumstances are relevant? Who is responsible for the solution? What else do I need to know?
2. **H**ARM—Who was (or would be) harmed by the situation or action? Who benefits (or would benefit)?
3. **E**VALUATION—What do I think about the situation? What are the options? What are the consequences of these options?
4. **A**CTION—What shall I do? What will be the consequences?
5. **P**REVENTION—How can situations like this be prevented?

The Evaluation step involves Kidder's tests for right vs. wrong:

- The legal test: Is it against the law or university regulations?
- Stench test: Does it smell?
- Frontpage test: What if everyone knows about the case?
- Mum test: What would my mum do?

It also involves Kidder's tests for right vs. right: Truth vs. loyalty, short-term vs. long-term, justice vs. mercy and self vs. community. The students need to understand that sometimes there are conflicts between positive values, and they should learn how to weigh them.

Session Outline

This part presents a newly developed session on ethical dilemmas, which has been part of the Teaching Lab course since the autumn of 2022. When referring to students, we mean the students of the Teaching Lab, i.e., student TAs.

The goals of the session were as follows: After the session, students should be:

1. able to deal with ethically difficult situations in teaching practice,
2. aware of potential moral dilemmas and know how to prevent them.

The session outline (details follow):

1. Check-in
2. Dyads
3. Discussion of selected issues
4. Ethical guidelines (incl. the CHEAP framework)
5. Analysis of case studies
6. Formal processes at the faculty + who to approach
7. Check-out

Check-in: This is the usual start of all Teaching Lab sessions. Students sit in a circle, and each of them answers a question. The question for this session is, *"Who is my role model teacher and why?"* It can be expected that some students value not only the professional skills of their (past) teachers but also fairness, trust and other ethical values.

Dyads: After check-in, the students make pairs and share an experience about a situation related to their teaching that they have had bad feelings about. One describes the case, and the other tries to analyse the situation and figure out what went wrong, what could be done better and how to prevent such situations. After 3 min, they switch their roles.

Discussion of selected issues: The following activity involves the whole group in a so-called "group thermometer". Within the activity, the group discusses several scenarios. The "thermometer" corresponds to the distribution of students' opinions in the classroom on how much the presented situation is an ethical problem. We define a scale and map it on one dimension of the room. The closer to the window, the more participants are convinced that a situation is an ethical problem. The closer to the door, the more they think it is not. After reading the scenario, the participants stand in a place in the classroom that represents their opinion. As soon as everyone knows their position, the discussion starts. First, we encourage the participants in extreme positions to share their reasons. Then, anyone can participate in the debate. During the discussion, participants are allowed (and encouraged) to change their position if they change their minds.

In the first run of the session, we picked the most controversial scenarios from the questionnaire:

1. The scenario with the largest variance: Using cocaine or other illegal drugs in your private (nonteaching) life;
2. The scenario with the highest number of respondents unable to judge: *Engaging in sexual fantasies about students*;
3. The scenario students provided within an open question: When assessing the assignments, consider how many points the student needs to pass.

In the next semester, we mainly discussed the situations on the edge between professional and private life, leading to conflict of interest. Apart from using drugs in private life and taking into account the number of points the student needs to pass, we also discussed:

- Taking advantage of a student's offer, such as wholesale prices at parents' store
- Inviting a student from my group to a date

Ethical guidelines: After the practical activities, the session continues with a 20-minute theoretical block on fundamental ethical theories, academic integrity values/virtues and brief guidelines for ethical decision-making. Within this block, students learn the CHEAP framework described above.

Analysis of case studies: Equipped with the theoretical background, students apply it to solve practical real-world dilemmas. The students split into 4–5 groups. Each group gets a description of a more complex moral dilemma and is asked to analyse the dilemma using the CHEAP framework. Then, each group shares their analysis with the others. The situations for the analysis are based on the real situations that happened at our Faculty. In some cases, we changed them to make them more difficult to judge:

- **Suspected plagiarism**: Upon checking students' homework, I noticed two students had turned in the same solution. Given the complexity of the assignment, this can't be a coincidence. I reported the case to the lecturer, who told me this was normal and I should not deal with it. What should I do?
- **A suspected leak from a colleague**: Lojzik was a good student who completed the course last year and is tutoring it for the first time this year. He has several classmates in his group who are repeating the course. And they are allegedly getting extra information from him about what tasks will appear on the exam. At least, that's what my students told me in an informal meeting. What should I do?

- **Unusual addressing**: I'm tutoring a subject for the first time, so I decided to visit an experienced colleague's class. I was intrigued by the way he called students to the board. He called one of them out with the words: "You don't seem to be able to count to five; let's try to figure it out." He said to another one, "Come on, you with the big crooked nose." When he then called the student up with the words, "That lady with the big tits is going to the blackboard over there," the classroom was abuzz. What should I do?
- **Misinformation in the course discussion forum**: A fellow tutor answers a student's question in the course discussion forum. Unfortunately, his answer is incorrect and reinforces a common misconception among students. I can immediately post the correct answer, but I don't want to shame him publicly. So, I will write to him privately and ask him to correct his answer. However, he is not responding to my messages, and I can see on Facebook that he is spending the whole weekend in the mountains. However, the students will be studying this weekend for a mid-term test on Monday. What should I do?
- **Individual deadline**: After a class, a student—one of the best in my group—comes to me and wants to hand in an assignment due 2 days ago. He says his grandfather died, and his mother had a breakdown, so he had a hard time at home. He′s been going to school (so he doesn't even have an excuse in the study system), but he couldn't do his homework on time. Without the points on the homework, he said he'd lose his chance to get an A and lose his merit scholarship. The assignments are on paper, and I have a stack of them at home. If I add one, no one will know. What shall I do?

Formal processes at the faculty + who to approach: After the group work, another theoretical block comes. Students get a brief overview of relevant policies and regulations at our university/faculty and learn about related processes and committees. They also learn about the dedicated people they may approach when they feel something is wrong or need help or advice.

Check-out: This is traditionally the last activity of each Teaching Lab session. Students sit in a circle, and each share the most important piece of knowledge they have learned. This triggers an immediate reflection of the finishing session (reinforcing learning) and provides valuable feedback for the teachers. In the semesters of autumn 2022 and spring 2023, the check-out also included a collection of suggestions on what should be included in the Code of Ethics for Tutors.

Code of Ethics for Tutors

As mentioned above, most of the seminars at FI MU are taught by student TAs. These TAs are both students and employees of the Faculty. Therefore, they are in an inherent conflict of interest. Several senior lecturers at the Faculty felt the necessity for explicit guidelines for student TAs, but there was no formal requirement for such a document.

The first draft of the ethical guidelines for student TAs was elaborated by the first author of this chapter already in the spring semester of 2022. The document contained comments on relevant points of Masaryk University's Code of Ethics. Although it was an essential first step, several colleagues were not happy with the document, as it did not cover several aspects necessary for the role of *student* teaching assistants. Therefore, we chose a more systemic approach and decided to elaborate the document based on the questionnaire and Teaching Lab sessions described in the previous sections.

We drafted the second version at the end of the autumn semester 2022. This version already contained specific guidelines tailored to the problematic situations. We identified the problematic situations from the questionnaire (primarily from open-ended questions) and in-person discussions with the Teaching Lab students. In the spring semester of 2023, at the Teaching Lab session on ethical dilemmas, we dedicated significantly more time to collecting issues the students wanted to include in the ethical guidelines. This further enriched the document. Over the summer break of 2023, we further worked on the document wording and structure. We titled the document Code of Ethics for Tutors. In our context, the term Tutor is more general and includes both student TAs and full-time Faculty employees. We circulated the draft to all FI MU teachers and asked them for feedback. Several colleagues provided helpful comments, which led to further improvements to the document.

On the 16th of February 2024, the Academic Senate of FI MU approved the document. Although it is just a declaratory document without any legal power, it expresses the will of the FI MU academic community and sets the standard for the tutors' ethical conduct. The Code is available from the FI MU website (Faculty of Informatics, Masaryk University, n.d.), linked from several places.

We would like to emphasise several aspects of the process that ensured wide acceptance of the Code:

- **Bottom-up approach**: The initiative did not come from the faculty management but from "ordinary" faculty teachers.
- **Involvement of the Teaching Lab community**: Teaching Lab is generally considered a high-quality course, and its teachers are respected for their pedagogical skills. This gave an informal authority to draft the Code.
- **Methodological approach**: The collection of issues from the questionnaire, disciplinary committee and Teaching Lab sessions contributed to the high quality of the draft.
- **Involvement of the academic community**: All FI MU teachers had the opportunity to provide their feedback on the draft proposal.
- **Official approval**: The Academic Senate, i.e., the academic community representatives, approved the Code.

The Code, as created and approved now, already fulfils an important function as a public statement and an internal reference document of academic integrity. However, this is only the start of a long journey: The following tasks are to make the document used, understood and regularly updated. The Faculty is also considering making it part of the contracts between the faculty and TAs to make it an unavoidable part of the teaching status. To help the target group understand individual clauses of the code better, we could extend each ethical principle presented in the Code with a specific example of ethical and unethical behaviour.

Conclusion and Future Work

This chapter presented a multi-faceted journey towards more explicit ethics education for student teaching assistants at FI MU. We provided insight into the design of a teaching toolkit on ethical dilemmas, which has been incorporated into the Teaching Lab course. We shared the results of a questionnaire-based survey conducted among the wide Teaching Lab community, which allowed us to understand the student TA's ethical beliefs and controversies and design effective training activities. Last but not least, we shared our experience with drafting a Code of Ethics for Tutors, which is unique in the Czech Republic and aims to ensure a high standard of ethical conduct not only for student teaching assistants, but also for all tutors at the Faculty.

The chapter reports on an experience at a single faculty in one Czech university, and, as such, its direct transferability to other cultural contexts is limited. Despite this hurdle, we believe the chapter provides valuable inspiration for those looking to address similar issues and implement related activities within their own institutions.

Appendix: Survey Instrument

Data about you

For how many semesters have you been involved in teaching at FI MU?
In what ways have you been involved in teaching at FI MU?
Your current relationship with the university

Ethical Assessment of Teaching-Related Behaviours

How ethical do you believe the following teaching and teaching-related behaviours to be?
(Definitely is an ethical problem/Probably is an ethical problem/Neutral/Probably is not an ethical problem/Definitely is not an ethical problem/I cannot judge)

1. Ignoring strong evidence of cheating
2. Giving easy courses or tests to ensure your popularity with students
3. Dating a student

4. Asking for small favours (e.g., a ride home) from students
5. Hugging a student
6. Accepting a student's inexpensive gift (worth less than $5)
7. Teaching when too distressed to be effective
8. Becoming sexually involved with a student
9. Accepting a student's expensive gift
10. Lending money to a student
11. Accepting a student's invitation to a party
12. Selling goods (e.g., textbooks or your car) to a student
13. Being sexually attracted to a student
14. Teaching a class without adequate preparation that day
15. Teaching material you haven't really mastered
16. Engaging in sexual fantasies about students
17. Encouraging students to participate in your research projects
18. Teaching while under the influence of alcohol
19. Teaching in laboratories with insufficient equipment/resources for all students
20. Using a grading procedure that does not adequately measure what students have learned
21. Teaching in classes so crowded you couldn't teach effectively
22. Teaching while under the influence of cocaine or other illegal drugs
23. Allowing a student's "likability" to influence your grading
24. Using profanity in lectures
25. Teaching content in a non-objective or incomplete manner
26. Encouraging competition between students
27. Engaging in a sexual relationship with a professor or other faculty member in your department
28. Allowing students to drop courses for reasons not officially approved
29. Inadequately supervising students in labs or classrooms
30. Grading on a strict curve regardless of class performance level
31. Using films and so forth to fill class time (and reduce your teaching work) without regard for their educational value
32. Teaching personal ethics or values to students
33. Telling colleagues confidential disclosures told to you by a student
34. Failing to update lecture notes when reteaching a course
35. Doing as instructed by a professor, regardless of your disagreement or ethical concern
36. Privately tutoring students in the department for a fee
37. Shortening or cancelling office hours without notice
38. Criticising all theoretical orientations except those you personally prefer
39. Taking advantage of a student's offer such as wholesale prices at parents' store
40. Insulting, ridiculing, and so forth a student in the student's presence
41. Ignoring the unethical behaviour of other teaching assistants
42. Using cocaine or other illegal drugs in your private (nonteaching) life
43. Ignoring the unethical behaviour of professors or faculty

44. Becoming sexually involved with a student only after he or she has completed your course
45. Teaching a course outside of your speciality (e.g., medical student teaching economics)
46. Teaching without experience or training
47. Teaching without training in ethics
48. Criticizing the teaching methods of colleagues in front of the students
49. Insulting, ridiculing, and so forth a student in the student's absence
50. Leaking confidential information about the upcoming exam to students in their seminar group

Additional Information

What question(s) did you hesitate the most?
Can you think of any ethically problematic situation that you got into as a teacher?

References

Branstetter, S. A., & Handelsman, M. M. (2000). Graduate teaching assistants: Ethical training, beliefs, and practices. *Ethics & Behavior, 10*(1), 27–50. https://doi.org/10.1207/S15327019EB1001_3

Bretag, T., Mahmud, S., Wallace, M., Walker, R., McGowan, U., East, J., Green, M., Partridge, L., & James, C. (2014). 'Teach us how to do it properly!' An Australian academic integrity student survey. *Studies in Higher Education, 39*, 1150–1169. https://doi.org/10.1080/03075079.2013.777406

Decker, A., Ventura, P., & Egert, C. A. (2006). Through the looking glass: Reflections on using undergraduate teaching assistants in CS1. In *Proceedings of the 37th SIGCE technical symposium on computer science education* (pp. 46–50). https://doi.org/10.1145/1121341.1121358.

Drumwright, M., Prentice, R., & Biasucci, C. (2015). Behavioral ethics and teaching ethical decision making. *Decision Sciences Journal of Innovative Education, 13*(3), 431–458. https://doi.org/10.1111/dsji.12071

Erdei, R., Springer, J. A., & Whittinghill, D. (2017). An impact comparison of two instructional scaffolding strategies employed in our programming laboratories: Employment of a supplemental teaching assistant versus employment of the pair programming methodology. *Proceedings of the IEEE Frontiers in Education Conference.* https://doi.org/10.1109/fie.2017.8190650

Faculty of Informatics, Masaryk University. (n.d.). *Code of ethics for tutors*. https://www.fi.muni.cz/about/code-of-ethics-for-tutors.html.en

Foltýnek, T., & Dlabolová, D. (2020). Academic integrity in Eastern Europe: Beyond corruption and plagiarism. In T. Bretag (Ed.), *A research agenda for academic integrity* (pp. 40–54). Edward Elgar. https://doi.org/10.4337/9781789903775

Foltýnek, T., Krásničan, V., & Henek Dlabolová, D. (2024). Academic integrity in Czechia and Slovakia. In S. E. Eaton (Ed.), *Second handbook of academic integrity* (pp. 135–153). Springer Nature Switzerland. https://doi.org/10.1007/978-3-031-54144-5_86

Hofstede, G. (2001). *Culture's consequences: Comparing values, behaviors, institutions and organizations across nations*. Sage Publications.

Kidder, R. (2009). *How good people make tough choices: Resolving the dilemmas of ethical living* (Rev. ed.). Harper Perennial.

Mahmud, S., Bretag, T., & Foltýnek, T. (2018). Students' perceptions of plagiarism policy in higher education: A comparison of the United Kingdom, Czechia, Poland and Romania. *Journal of Academic Ethics, 17*, 271–289. https://doi.org/10.1007/s10805-018-9319-0

Osipian, A. L. (2017). University autonomy in Ukraine: Higher education corruption and the state. *Communist and Post-Communist Studies, 50*(3), 233–243. https://doi.org/10.1016/j.postcomstud.2017.06.004

Perri, D. F., Callanan, G. A., Rotenberry, P. F., & Oehlers, P. F. (2009). Education and training in ethical decision making: Comparing context and orientation. *Education + Training, 51*(1), 70–83. https://doi.org/10.1108/00400910910931841

Pivkina, I. (2016). Peer learning assistants in undergraduate computer science courses. *Proceedings of the IEEE Frontiers in Education Conference*. https://doi.org/10.1109/fie.2016.7757658.

Ren, Y., Krishnamurthi, S., & Fisler, K. (2019). What help do students seek in TA office hours? In *Proceedings of the 2019 ACM Conference on International Computing Education Research Workshop*. https://doi.org/10.1145/3291279.3339418.

Riese, E., & Kann, V. (2020). Teaching assistants' experiences of tutoring and assessing in computer science education. In *Proceedings of the 2020* IEEE Frontiers in Education Conference, Sweden. https://doi.org/10.1109/fie44824.2020.9274245.

Riese, E., Lorås, M., Ukrop, M., & Effenberger, T. (2021). Challenges faced by teaching assistants in computer science education across Europe. In *Proceedings of the 26th ACM Conference on Innovation and Technology in Computer Science Education*. https://doi.org/10.1145/3430665.3456304.

Stöckelová, T., & Virtová, T. (2015). A tool for learning or a tool for cheating? The many-sided effects of a participatory student website in mass higher education. *British Journal of Educational Technology, 46*(3), 597–607. https://doi.org/10.1111/bjet.12155

The Embassy of Good Science. (n.d.). *02 - The seven steps method: A method for analysing cases in research ethics and research integrity*. https://embassy.science/wiki/Instruction:6b129846-c455-4849-9eaf-0d25f3c5600e

Ukrop, M. (2022). *Two cases of human-centered computer science: Usable developer interaction with TLS certificates and effective teaching assistant training* [Doctoral dissertation, Masaryk University]. Masaryk University. https://is.muni.cz/th/wwx5a/PhD_thesis.pdf

Ukrop, M., Švábenský, V., & Nagy, I. (2020). Teaching lab: Training novice computer science teachers. In *Proceedings of the 2020* ACM Conference on Innovation and Technology in Computer Science Education (ITiCSE '20). Association for Computing Machinery. https://doi.org/10.1145/3341525.3393967.

Vlková, I. (2019). Minimum pedagogical education for university teachers. *EDULEARN19 Proceedings*, 7813–7818. https://doi.org/10.21125/edulearn.2019.1899

Wotruba, T. R. (1993). A framework for teaching ethical decision making in marketing. *Marketing Education Review, 3*(2), 2–8. https://doi.org/10.1080/10528008.1993.11488404

Chapter 10
Countering Education Fraud and Essay Mills Through Legal Measures

Michael Draper

Abstract This chapter discusses the 2022 Recommendation on Countering Education Fraud, adopted by the Committee of Ministers of the Council of Europe, which addresses various forms of education fraud, including the more complex operations of accreditation mills, essay mills, and diploma mills. The Recommendation was developed by the Council of Europe Platform on Ethics, Transparency and Integrity in Education (ETINED), a network of specialists from 50 member States dedicated to promoting ethical principles in education. ETINED's mission includes sharing best practices, defining guidelines, and building capacity to foster transparency and integrity in education through a principled approach, advocating for societal commitment to ethical standards rather than relying solely on regulatory measures. The Recommendation emphasises the importance of international cooperation and legal frameworks to combat education fraud, recognising its detrimental impact on public safety, educational quality, and equal access to opportunities. The chapter details the extensive research and collaborative efforts undertaken by ETINED to draft the Recommendation, highlighting the need for a common framework that respects the diverse legal, cultural, and technological contexts of member States whilst promoting academic integrity through a participatory approach. A review of the initial impact after introducing legislation in England to address essay mills is included as a case study of how national legislation can support the 2022 Recommendation. The chapter identifies the importance of a balanced approach between national legislation and international cooperation to effectively address cross-border education fraud to provide a secure foundation to enhance academic integrity and combat education fraud across Europe.

Keywords Education fraud · Essay mills · Legal measures · Legislation

M. Draper (✉)
Swansea University, Swansea, UK
e-mail: m.j.draper@swansea.ac.uk

I. Glendinning et al. (eds.), *Ethics and Integrity in Education (Research)*, Ethics and Integrity in Educational Contexts 9,
https://doi.org/10.1007/978-3-031-98157-9_10

Introduction

The Council of Europe Platform on Ethics, Transparency and Integrity in Education (ETINED) is a network of specialists appointed by member States of the Council of Europe and of States Parties to the European Cultural Convention (50 States). Its declared mission is threefold:

- to share good practices in the field of transparency and integrity in education,
- to define guidelines on the subject, and
- to develop capacity-building for all actors.

ETINED seeks to build a principled based approach on the foundation that quality education will only be achieved and corruption effectively addressed if all relevant sectors of society commit to fundamental positive ethical principles for public and professional life, whilst not relying solely upon top-down regulatory measures. Although not the only means for doing so, legal norms and measures are important in supporting the foundational principles of ethics, transparency, and integrity. In July 2022, the Committee of Ministers adopted the Recommendation on Countering Education Fraud (Council of Europe, 2022a) with the accompanying Explanatory Memorandum (Council of Europe, 2022b) noting worldwide concern over education fraud at all levels of education ranging from plagiarism to the criminal offences of impersonation or identity theft and organisations promoting essay mills (otherwise known as "contract cheating" or "academic custom writing services"), visa mills, accreditation mills and diploma mills.

The 2022 Recommendation represents the collective will of member States to tackle serious issues affecting academic integrity including the authenticity of qualifications, international reputation and risks to society as a whole through issues relating to public safety. The ETINED principles are a primary reference from which the 2022 Recommendation emerges and represents an evolution of other measures including Recommendation CM/Rec(2012)13 on ensuring quality education (Council of Europe, 2012). This asserts that corruption in education constitutes a violation of the basic principle of equal rights and calls for national legislation to address these issues.

The 2022 Recommendation is part of a continuum of measures designed to address corruption and education fraud and promote ethics, transparency, and integrity in education, as an example please see Recommendation CM/Rec (2014) of the Committee of Ministers to member States on the protection of whistle-blowers (Council of Europe, 2014).

The 2022 Recommendation on Countering Education Fraud

The 2022 Recommendation (Council of Europe, 2022a) has four key themes: prevention, prosecution, international cooperation, and monitoring, expressed through 17 articles that identify six core measures to tackle education fraud and to promote ethics, transparency and integrity in education. It was recognised early on in the drafting of the Recommendation that the promotion of these values was essential to create a comprehensive international legal instrument that was relevant to the national settings of member States.

Over 4 years of extensive research and analysis were conducted by the ETINED platform representatives to understand the various forms of education fraud, including diploma mills, accreditation mills, and essay mills. This phase involved gathering data from member States and identifying best practices and challenges applicable to individual state settings through a series of targeted questionnaires to member States. Collected data was presented to all delegates and discussed in plenary sessions of the ETINED platform, which occurred each year in Strasbourg. It was essential to identify the various forms of education fraud operating in each member state and establish agreed definitions around education fraud before commencement of the drafting of a legal measure. It was also important to acknowledge that member States have differing priorities in relation to their national educational settings and reflect these differences in drafting with appropriate guidelines for implementation. Agreement was reached by the second year, and a small working group of six ETINED expert members was formed to commence drafting based on the collected data and resulting guidance derived from the plenary discussions. The author was a member of the drafting group and was subsequently given the responsibility of presenting working drafts to delegates at further plenary sessions and in other exploratory sessions in European capitals. The draft recommendation, therefore, underwent several rounds of review, feedback and refinement, and this iterative and participatory process ensured that the document was robust and addressed the concerns and suggestions of member States so as to facilitate ultimate adoption. The final draft was then submitted to the Steering Committee for Education Policy and Practice (CDPPE) for approval. The CDPPE is responsible for overseeing the development and implementation of education policies within the Council of Europe, with the author delivering the presentation in session to them, which was approved without amendment; a recognition of the successful participatory and inclusive approach taken during the developmental phase of drafting.

The ETINED approach to education fraud and corruption is influenced by a complex interplay of legal, cultural, technological, and resource-related factors unique to each member state. Each member state has its own legal framework, with a minority adopting laws and regulations specifically addressing education fraud, while others have more general anti-corruption legislation. The structure and organisation of education systems vary significantly, with differences in governance, curriculum, and assessment methods impacting on the implementation of fraud prevention measures in circumstances in which cultural norms and attitudes toward corruption are varied. Member States also differ in their adoption of technology in the delivery of education.

The 2022 Recommendation recognises this diversity within a common framework of core measures that support:

- the promotion of quality education by eliminating education fraud;
- the protection of pupils, students, researchers, and staff at all levels of education from organisations and individuals engaged in selling (and advertising) fraudulent services;
- the implementation of preventative and protective measures, as well as a culture of equality of opportunity at all levels and in all sectors of education and training and in the transition between these sectors;
- the monitoring of technological developments that could support new forms of fraud;
- international cooperation, and
- wide dissemination of the Recommendation.

(adapted from Council of Europe, 2022a, pp. 7–8)

The scope of the 2022 Recommendation is broad and inclusive, with the measures intended to apply to all levels and forms of education, including pre-primary to higher education, and vocational education and lifelong learning, as well as access into those levels. A common understanding is critical to achieve the aims and objectives, and, therefore, at the outset, a detailed set of 11 definitions relating to Codes of Ethics and fraudulent services, including diploma mills and essay mills, is to be found. The need for clarity around terminology served as the impetus for development of the Glossary for Academic Integrity developed by the European Network for Academic Integrity (Tauginienė et al., 2019).

Codes of Ethics have an important role as a recognisable framework tool in the ETINED platform expressing values and principles that inform aspirations and responses within national settings and serve to develop and maintain ethical behaviour and transparency to support and promote academic integrity (Council of Europe, 2019). They are sensitive to the policy environment of each member state and are designed to be accessible and relevant for all stakeholders, with review and development as the education fraud environment changes. A participatory approach to development and review in consultation with relevant stakeholders is important to establish ownership and responsibility for the standards set out in each Code. This participatory approach ensures they are embedded in educational culture and training, but successful implementation requires a broad and sustained campaign of awareness raising. Consequently, within a framework of common principles, Codes may be adapted to meet the needs and position of member States.

Consistent with the siting within a continuum of measures, Codes of Ethics align with the Reference Framework of Competences for Democratic Culture (RFCDC) (Council of Europe, 2013), another major initiative of the Council of Europe. The RFCDC aims to provide a systematic approach in designing the teaching, learning, and assessment of competences for democratic culture, with a list of 20 competences to empower learners to act as competent and effective democratic citizens.

Education fraud corrupts societies and defrauds organisations and individuals. Many fraudulent education service providers operate (or claim to operate) from member States, which gives the community a unique position to address this threat to societal and educational values. In this context, awareness raising actions through Codes of Ethics should be partnered with effective legal action through the enforcement and review of national legislation.

Legal Frameworks Laws and Practices

The 2022 Recommendation incorporates two express provisions relating to a legal response to education fraud:

6. Advertising and promotion of education fraud

Member States should, in accordance with national legislation, take steps to prohibit the advertising and promotion of fraudulent education services, whether in printed form or via online media, including through social media and the internet.

7. Legal frameworks, laws and practices

Member States should take all necessary and appropriate action to use existing legislation, guidelines or practices to eradicate education fraud and the activities of fraudulent education service providers. They should also consider introducing new legislation or policy measures where required and encourage all education institutions to adopt regulations consistent with that aim. In doing so, member States and education or training organisations should take appropriate steps to protect the rights of pupils, students, researchers and staff.

(Council of Europe, 2022a, p. 12)

These articles are designed to operate within national settings and do not necessarily require additional measures to be adopted. Albeit member States are asked to consider introducing new legislation or legal policy and required to ensure the effectiveness of existing legal measures to counter education fraud. It is recognised and commended by the ETINED Platform that some member States have adopted legislation which either criminalises or otherwise provides for legal redress against certain education fraud activities, albeit the enforcement of legislation, including detecting and prosecuting offenders, is challenging. This recognition was necessary to ensure the adoption of the 2022 Recommendation as a working legal instrument informing national conversations and developments. Member States have the flexibility to find the right balance between legal and other measures in accordance with their national contexts.

Fraudulent education service providers (as outlined in Article 2 of the 2022 Recommendation, pp. 8–10) are essentially business entities with the primary objective of generating profit. As with all businesses, they rely heavily on advertising and marketing strategies to boost sales and attract "customers". These fraudulent providers utilise various channels and tools, including social media, to reach potential clients (for example students) and market their products.

While education fraud is not a new phenomenon, social media and new technologies have enabled these fraudulent providers to offer their services on a global scale (Draper & Crossman, 2019). In situations where it is not possible to take direct action to close down the provider (for example, because it is not located in the member state or its activities are not deemed illegal under national law), member States can still take steps to reduce education fraud and combat the supply side of the issue by undermining the effectiveness of the providers' marketing strategies.

Possible actions include promoting guidelines and policies for education providers, which require them to indicate on their website and on the certificate itself if the qualification awarded has no legal value in the awarding country; reporting unfair and opaque practices to the competition authority or the body responsible for consumer protection in the country, which could take action against misleading advertising. In countries where relevant education terminology is protected (Article 9), member States could take action against providers that provide misleading information, and for example, as in the case of England (which we will consider later) prohibiting and banning the advertising of essay mills and providers of contract cheating; and promoting initiatives to raise awareness among students and potential customers of such services.

To give effect to Article 7, member States need to make the establishment, licensing, and operation of fraudulent education service providers within their jurisdictions illegal and prosecute these entities when possible. Furthermore, member States should restrict or ban the use of fraudulent credentials for employment and promotion. This can be done either by enforcing existing education, consumer and immigration laws, and/or the introduction of new legislation. ETINED recognises that there are several tools available to combat educational fraud, including informing the public about the harm caused by fraudulent credentials. This article encourages member States to identify and use the tools that fit their national context.

Essay Mills were viewed as a particular threat and have their own definition in Article 2 of the 2022 Recommendation as an organisation or an individual entering into contracts with students to complete, in full or in part, one or more assignments (including student work such as essays, projects, theses and dissertations) for financial gain. Essay Mills are predominantly online entities with a global reach, and students who use them are increasingly at risk of extortion and blackmail (Draper & Crossman, 2019). These services often store personal data online with little to no security measures in place, leaving students vulnerable to identity theft and bank fraud.

The next section describes a case study how the national legislation for England on Essay Mills was developed in support of the Council of Europe Recommendation some 3 years after Ireland introduced their legislation targeting commercial contract cheating services through section 43A of the Qualifications and Quality Assurance (Education and Training) (Amendment) Act 2019 which made it an offence to facilitate a learner to cheat; advertise cheating services to leaners and publish advertisements for cheating services to learners. Whilst the author was in contact with Quality and Qualifications Ireland prior to the Irish legislation being introduced to advise on the draft provisions it is to be noted that the Irish legislation does not adopt the principle of 'strict liability' which we will discuss in the next section.

Case Study: Legislation Against Essay Mills in England

In parallel with the drafting and adoption of the 2022 Council of Europe Recommendation on countering education fraud, the arguments in favour of targeted legislation in relation to Essay Mills eventually met with success in the UK. Initially proposed as a course of action in 2016/17 (Draper & Newton, 2017), on 28th April 2022, some 3 months prior to the adoption of the 2022 Recommendation, sections 26–30 of the Skills and Post 16 Education Act 2022 (UK Government, 2022) were brought into force in England.

Chapter 1 Part 4 of the 2022 Act, applying to education in England, provides for new offences that criminalise individuals and organisations who provide or arrange for another person to provide (in commercial circumstances) contract cheating services for students at post-16 institutions and any other person over compulsory school age who has been entered to take an examination relating to a regulated qualification at a place in England. The provisions also criminalise the advertising by those individuals or organisations of the services.

These offences directly align with the intent of Articles 6 and 7 of the 2022 Recommendation (Council of Europe, 2022a, p. 12), and it is notable that the offences are not limited to activity in relation to higher education but extend to contract cheating services for students at all post-16 educational institutions, which brings sixth forms and further education colleges into scope.

There are two principal offences in the English Skills and Post 16 Education Act 2022:

- Section 27: Offence of providing or arranging a relevant service
- Section 28: Offence of advertising a relevant service

Section 27 provides that it is an offence for a person to provide or arrange for another person to provide, in commercial circumstances, contract cheating services for a student in relation to a relevant assignment. In order for an individual or body to be liable under the offence, they must complete all or some of the student's work such that the work would not be considered the work of the student. This excludes generally accepted study support such as tutors, proof-reading, and ordinary teaching practices from being covered by the offence, as even with the provision of such services, the work would generally still be considered to have been completed by the student personally.

Section 28 provides that an offence is committed if a person advertises a relevant service to students. The advertisement does not need to be advertised solely to its target audience and might not be seen by its target demographic, but the offence lies in the simple fact of advertising the relevant service. Advertising is not defined in the Act, so any promotional activity could potentially be covered, for example, via

WhatsApp; students frequently complain that they receive multiple unsolicited promotional messages on a regular basis.

There are two key points to appreciate. Firstly, these are strict liability offences. That is, they are satisfied simply by establishing the fact of supply or advertisement. This is a significant advantage to prosecution as the evidential burden and cost of proving the contrary is on the defendant's essay mill, that is, that they did not know and could not, with reasonable diligence, have known of the relevant matters. So, for example, the defendant's essay mill may need to be able to prove that they did not know, and could not with reasonable diligence have known, that the student was required to complete this work personally. Similarly, they might need to prove that they did not know and could not, with reasonable diligence, have known that the service provided was not specifically permitted. This is significant because essay mills typically rely on disclaimers around use in their terms and conditions, despite the fact that it is generally known through publicity, the media and otherwise that students use these services to cheat. A successful defence would require more than simple reliance on terms and conditions of supply. As noted above, the Irish Qualifications and Quality Assurance (Education and Training) (Amendment) Act 2019 does not make use of this principle and nor does the Council of Europe 2022 Recommendation require it. The principle of 'strict liability' is a significant enhancement for the reasons discussed above.

Secondly, it is specifically provided that a student who commissions a provider to complete work will not be liable for an offence of encouraging, assisting, aiding or abetting the substantive offence. Importantly, students are not criminalised by this activity. Interestingly, the 2022 Recommendation does not specifically address this point, but the ETINED platform has, from the authors' experience, in plenary sessions leading up to the adoption of the 2022 Recommendation, discussed a similar approach to avoiding student criminalisation with approval.

This has attracted international attention. On the 6th of December 2022, the author received an email from a policy advisor based at a Texan University which addressed some key points of general importance.

The policy advisor stated that he had been reviewing the evolution of the essay mill legislation, which appeared to him was at least in part based on the model that the author had proposed. He noted there was a group in the United States (unspecified) considering drafting legislation with a similar purpose and was seeking advice on a number of issues including the principle of strict liability. I advised that adopting the principle of strict liability avoided the problem of prosecutors having to prove intent or knowledge on the part of the supplier, so that the supplier or advertiser would commit an offence by simply demonstrating supply or an advert, subject to a defence of due diligence. I also considered it important to ensure that human agents of a corporation should not be able to avoid criminal liability by hiding behind the immunity given by the corporate veil of incorporation, and so the English legislation specifically addressed this point.

Furthermore, section 26(2) of the 2022 Act incorporates a targeted definition of "relevant service" so that in order for an individual or body to be liable under the offence, they must complete all or some of the student's work such that the work would not be considered the work of the student. This is important as it excludes generally accepted study support such as tutors, proof-reading and ordinary teaching practices from being covered by the offence, as even with the provision of such services, the work should generally still be considered to have been completed by the student personally.

However, section 26(4)(a) of the 2002 Act creates a presumption that where a student seeks the provision of a "relevant service," material provided as a result will be viewed as being provided in connection with the assignment (which is relevant for the purposes of subsection (3)). This means that, for example, where a student chooses an essay from a menu of available stock essays, this would be caught (provided the essays are not published generally (see subsection (4)(b)). So, the provider need not in fact know about the assignment or what it is in order to meet the test in subsection (3).

Therefore, in order for the offence to be committed, the relevant service must be provided, or arranged, "in commercial circumstances." This covers: an organisation or individual which is providing or arranging the cheating services in the course of their business; and an individual who works independently and has been commissioned to provide the cheating services, by a person acting in the course of their own business or by a person acting in the course of their employer's business. This definition allows for freelancers to be captured as well as employees and owners of companies providing cheating services.

Whilst every effort was made to avoid institutional and student involvement in relation to the commission of an offence, it is accepted that establishing the fact of supply is still required, so evidence is required from a student of this voluntarily, but it cannot be compelled. A student can take comfort from the fact that even though evidence is given, a student will not be guilty of an offence based on the act of purchase. Subsection 27(6) of the 2022 Act makes clear that a student who commissions a provider to complete work in a way that falls under the offence will not be liable for an offence of encouraging, assisting, aiding or abetting the substantive offence.

The reality is that most referrals for prosecution will be based on institutional academic misconduct proceedings. It is hoped that students, as part of that process and wider discussions with their institution, will volunteer relevant information about purchase from a specific provider.

The section 28 offence on advertising has the potential for greater impact because section 28 (2) of the 2022 Act makes clear that, in order for the offence to be committed, the advertisement does not need to be advertised solely to its target audience. The advertisement might not even be seen by its target demographic, but the offence lies simply in the fact of advertising the relevant service. Another example of the operation of the principle of strict liability in the 2022 Act.

Intent has to be proven for a successful prosecution under the Irish Qualifications and Quality Assurance (Education and Training) (Amendment) Act 2019 in relation to commercial contract cheating. We have explored above how the English legislation differs and enhances this position through the use of the principle of 'strict liability,' but the key question to be answered is whether the 2022 Act has had a practical impact on the operation of commercial contract cheating providers in England. This will be considered next.

Impact of the 2022 Act (England)

At the time of writing, there have been no prosecutions in relation to offences set out in the English Act.

This is not surprising. In reviewing the proposed law, two members of the UK QAA Academic Integrity Advisory Group (Crossman & Draper, 2021) noted that legislation a) will never be perfect and b) will never solve a problem alone. The reality is that most essay mills are based outside the UK, so would not be subject to domestic law. Even with UK based essay mills, it is likely that only a few would ever be prosecuted. Furthermore, unlike legislation in Ireland and Australia, prosecution powers do not sit with the regulator. The offences are free standing, so it would be for the police and CPS to initiate any prosecution. Realistically, it is unlikely that wholesale prosecutions of essay mills will ever be top of the policing agenda.

However, the view was held that the 2022 Act would make a difference because the offences are one of limited strict liability. A strict liability offence is one where the prosecution does not need to prove intent to gain conviction, and this is highly unusual in UK criminal law, where intention is a standard element of any offence.

An aim of the legislation was to encourage a change in student attitudes in relation to cheating behaviours and the nature of the conversation that their tutors might have with them on the importance of academic integrity (THE, 2022).

What is welcome is that the Essay Mills in the UK did change their behaviour in response to the legislation. In November 2022, on a routine search of one UK registered essay mill site (deliberately not disclosed), a pop up appeared:

"Confirm Country of Study

Your IP suggests that you may be based in England. It is not illegal to buy custom writing services in the UK, but it is illegal for us to sell to anyone studying at an English higher education institution.

Your order will not be processed if it breaches any of the criteria found on this page: click here.

Please ensure you have read and understood this page before placing your order if you are based in England.

If you believe you are seeing this message in error, it may be due to you using a VPN, such as Nord or PrivadopVPNFree, which is making your IP address one that is located in England. If so, please switch off your VPN and now refresh your order."

In July 2023, desktop research was undertaken to update research conducted in 2019 (Draper & Reid-Hutchings, 2019) on the operation of UK registered essay mill companies. The 2019 research was used to present arguments for legislation, and the update was undertaken to determine if the legislation might have had an impact on the number of UK registered essay mill companies.

In July 2019, there were twenty-six unique websites operated by twenty-one companies listed at UK Companies House. At the time of analysis in July 2023, twenty-one websites owned by eighteen companies were capable of being assessed. On investigation in 2023, only three UK based essay mill companies remained functional; this denotes a loss of about 67% of previously identified corporations from 2019.

It cannot be said with certainty that this is due to the legal environment created by the legislation, but it is significant as it demonstrates a distinct shift in the industry in the UK following the adoption of the criminal offences in the 2022 Act. Some essay mill companies have removed themselves from the UK in terms of legal incorporation (as opposed to advertising of services); it should be noted that: enforcement against a non-UK registered company is simply more challenging.

Nine companies have been liquidated since the Skills & Post 16 Education in England Act was introduced into Parliament or since the legal bringing into force of the 2022 Act (6). This may be due to the changing legal environment.

In the previous 2019 assessment, one limited company championed four independent websites offering essay writing services. Though the original limited company is still active and operational as of July 2023 (according to the UK companies house and its namesake website), its other former websites have since become registered under a new company located in Cyprus.

One remaining company operating out of Wales makes distinct reference to the 2022 Act in several places. As one of the remaining operational websites since the original 2019 assessment, this is clear evidence of the legislation's impact. The company has updated all relevant disclaimers to make explicit mention of the legislation and restrictions imposed on students in England and their ability to access services. Moreover, the terms and conditions now go beyond the general classification of the UK. It specifically references England as a restricted region; they include their own distinct paragraph outlining the parameters of the legislation and the relevance to the services offered. Thus, services are insistently for reference only.

The 2023 desk top research is summarised in Table 10.1.

Of the remaining three companies, their advertising at the time of investigation remains a concern, with that advertising considered misleading according to metrics identified by the UK Advertising Standards Authority (Table 10.2).

Despite the fact that there have been no prosecutions to date as demonstrated, the 2022 Act has had an impact on the environment in which commercial contact cheating companies operate at least with respect those companies taking advantage of the marketing advantage of being able to claim they are a UK registered company. What will be considered in the next section is whether the 2022 Recommendation is able to replicate the early success of an English legislative response to commercial contact cheating.

Table 10.1 July 2023 Desktop Assessment of Registered UK companies

Question	# sites	%
How many companies have changed status since 2019?	14	67%
How many companies have migrated their business registration to different legal jurisdictions?	6	33%
How many companies have dissolved, entered liquidation, or become dormant since the first reading of the Skills & Post 16 Education in England Bill?	9	50%
How many Companies have dissolved, entered liquidation or become dormant since the ratification of the Skills & Post 16 Education in England Act 2022?	6	33%
Companies whose domain appears to have been purchased.	5	28%
Companies whose URL link is dangerous or inoperable.	8	44%
Companies whose domain has been purchased and continues to operate as an essay mill by a non-registered UK company.	1	6%
Operational companies that have moved to another country but whose terms and conditions or business model recognise the Skills & Post 16 Education in England Act 2022 or its effects.	1	6%
Operational UK based companies whose terms and conditions or business model acknowledge or recognise the Skills & Post 16 Education in England Act 2022 or its effects.	1	6%
Remaining UK based companies whose websites are active and functional for essay writing services.	2	11%
Total number of UK based operational companies registered at companies house.	3	

Questions considered about the status changes to companies and/or their business models since 2019.

Table 10.2 July 2023, Advertising

Question	# sites	%
Is there a disclaimer of liability in respect of use?	2	66%
Are there contractual terms and conditions relating to use?	3	100%
Are customers able to specify the quality (grade) of the work?	3	100%
Is the intellectual property in the work retained by the company?	3	100%
Does the work come with a 'plagiarism-free' guarantee?	2	66%
Does the site state that the products are to be used for 'research' only?	3	100%
Does the site offer 'qualified' or 'expert' writers?	3	100%
Is the advertising potentially misleading (compared to disclaimer / terms + conditions)?	3	100%

Questions asked of the different websites and the number/percentage of sites for which the answer was considered to be 'yes'

Impact of the 2022 Recommendation

An initial review of the impact of the Recommendation was undertaken as described in Council of Europe Platform on Ethics, Transparency and Integrity in Education—Volume 7—Means to counter education fraud—Legislation, practices and instruments (Council of Europe 2023). The research studies in this book analyse the situation in countering education fraud in some member States at the time when the Recommendation was adopted and give suggestions for consistent implementation of it in the future. The author presented on this publication at the 7th Plenary meeting of the ETINED platform in November 2023.

The 2022 Recommendation provides for the systematic collection and exchange of data between member States (Article 15) to stimulate and support international co-operation in the prevention of education fraud. A baseline survey questionnaire against which future activity and impact could be measured was sent to all members of the ETINED platform in May 2022 to collect comparable data on the issues covered by the recommendation and to assess the educational fraud practices of its member States.

In relation to Essay Mills and contract cheating three questions were addressed:

1. What legislation currently exists in member States?
2. What is the enforcement policy in relation to that legislation: its impact or effectiveness including enforcement across state borders?
3. What legislation is planned in member States?

As expected, there was a limited number of member States with laws directly targeting contract cheating and essay mills. Nineteen member States were aware of education fraud activity, including essay mills, either generally or specifically occurring within the member state, but only one member State, the UK, had at that time recorded an intent to introduce legislation addressing this form of fraud. It was noted that England had done so, but Wales, Scotland and Northern Ireland had, at the time of the response, not followed England's lead. It is appreciated that enacting legislation requires political will and time within the context of competing legislative priorities applicable in and appropriate to national settings, but no reasons were given by respondents for their position.

Given the commitment in the 2002 Recommendation to collection and exchange of data observatory on countering education fraud has been proposed with the intention that it acts as a research centre as well as a hub for monitoring, tracking and conducting analysis in the field of fraudulent activities in education. This proposal was discussed at the 7th Plenary meeting of the ETINED platform in November 2023 with strong support from delegates, noting that this aligned with the Council of Europe's Education Strategy 2024–2030, which seeks to encourage and strengthen the prevention and monitoring of education fraud through data collection and comparative analysis.

Currently, Essay Mills continue to operate on global basis, but the advent of new model generative artificial intelligence (Gen AI) platforms may see further contraction of essay mill organisations. In the short term, this is by no means certain because mills themselves use Gen AI to produce work at speed and then check for and correct errors or hallucinations. This is a marketing advantage over student produced work using Gen AI when students do not have the knowledge or understanding to identify inaccuracies, false references, or other hallucinations. This advantage will be eroded as the accuracy of Gen AI platforms improves. The costs of using essay mill services as opposed to a subscription to a Gen AI platform will also factor in the decline of traditional essay mills. A subscription to a Gen AI platform provides unlimited and multiple access and user rights.

Neither the 2022 Recommendation nor the 2022 Act expressly addresses the use of Gen AI. At the time of drafting the recommendation and legislation, this was not in the minds of the law makers, and, therefore, there is no reference to it in the official records and minutes of debates and plenary sessions, as both predate the advent of Generative AI in the public consciousness (October/November 2023).

Conclusion

Education fraud poses a significant challenge in Europe, affecting the integrity and quality of educational systems. In response, the Council of Europe has taken a crucial step in addressing education fraud by adopting the 2022 Recommendation, which is designed to be a living document subject to periodic review as circumstances change. This will ensure that the 2022 Recommendation remains relevant to member States according to their national settings and requirements, and the proposed observatory monitoring the impact of the 2022 Recommendation on education fraud will assist with this aim. Countering education fraud through legal measures can have a significant impact, as demonstrated by the research undertaken in 2023 in relation to the 2022 Act (England), which has likely resulted in the number of Essay Mills registered in the UK being reduced to three, with one of those based outside the jurisdictional reach of England. Arguably, the principle of strict liability has played a significant role in creating an environment that is not conducive to the operation of commercial contract cheating services and as a drafting principle, it has much to commend it. It should be noted that Australian legislation addressing commercial contract cheating services also adopts this principle, whereas the legislation in Ireland and New Zealand does not. However, Ireland has had significant success in leveraging their legislation to ensure the removal of advertising of relevant cheating services through agreements with social media and other platforms, as well as initiating positive culture change within the Institutional higher education environment. The ability of Irish and Australian regulators to initiate prosecutions is also to be envied. Furthermore, whilst both Irish and English legislation allow for fines to be imposed as a punishment for offences, the Irish legislation goes further to include the possibility of a five-year prison term. This has a significant deterrent effect, even if the prosecution may be initially challenging because of the need to prove 'intent' to supply as opposed to being able to rely on the principle of strict liability, which reverses the burden of proof.

The Council of Europe's 2022 Recommendation provides a robust framework for tackling education fraud, however, cross-border cooperation and technological advancements remain essential to safeguard the integrity of European education systems. While not explicitly mentioned in the 2022 Recommendation, the EU AI Act 2024 (Regulation EU 2024/1689 of the European Parliament and of the Council of 13 June 2024 laying down harmonised rules on artificial intelligence) will play a pivotal role in this through a risk based approach leveraging technological advancements for fraud detection, authentication, and verification and so enhance the effectiveness of anti-fraud measures provided there is a balance struck between technological advancements and privacy safeguards. It is critical that, in the context of the pace of change, the legislative landscape continually evolves to meet the challenges posed by education fraud through collaboration, innovation, and above all, the adoption and implementation of effective guiding ethical principles. It remains to be seen if the UK will follow Europe's lead in introducing legislation addressing Gen AI. The UK government have recently published a framework and guidance and outlined its approach to regulating Gen AI in the UK using a risk and opportunity-based approach. Rather than giving responsibility for AI governance to a central regulatory body, the UK government's proposals will allow existing regulators to take a tailored pro-active approach to the use of Gen AI with a view to creating proportionate and adaptable regulation whilst encouraging coordination between regulators. Time will determine the effectiveness and desirability of this approach.

Acknowledgements The author would like to acknowledge Elenor Morano, who undertook the initial research for Tables 10.1 and 10.2.

References

Council of Europe. (2012). *Recommendation CM/Rec(2012)13 of the Committee of Ministers to member States on ensuring quality education.* https://www.ecml.at/Portals/1/documents/CoE-documents/CMRec2012-13_quality_EN.pdf?ver=2016-11-29-113145-700

Council of Europe. (2013). *The reference framework of competences for democratic culture (RFCDC).* https://www.coe.int/en/web/reference-framework-of-competences-for-democratic-culture

Council of Europe. (2014). *Protection of whistleblowers - Recommendation CM/Rec(2014)7 and explanatory memorandum.* https://book.coe.int/en/legal-instruments/6162-protectionof-whistleblowers-recommendation-cmrec20177-and-explanatory-memorandum.html

Council of Europe. (2019). *Recommendation CM/Rec(2019)9 of the Committee of Ministers to member States on fostering a culture of ethics in the teaching profession.* https://search.coe.int/cm?i=090000168094fa96

Council of Europe. (2022a). *Recommendation CM/Rec(*2022a*)18 of the Committee of Ministers to member States on countering education fraud.* https://www.coe.int/en/web/ethics-transparency-integrity-in-education/-/new-recommendation-cm/rec-2022-18-and-explanatory-memorandum-countering-education-fraud

Council of Europe. (2022b). *Explanatory memorandum to CM/Rec(*2022b*)18.* https://search.coe.int/cm/Pages/result_details.aspx?ObjectId=0900001680a6e557

Council of Europe. (2023). *ETINED - Council of Europe platform on ethics, transparency and integrity in education – Volume 7: Means to counter education fraud – Legislation, practices and instruments.* https://book.coe.int/en/education-policy/11748-etined-council-of-europe-platform-on-ethics-transparency-and-integrity-in-education-volume-7-means-to-counter-education-fraud-legislation-practices-and-instruments.html

Crossman, G., & Draper, M. (2021, December 2). *Criminalising essay mills will make a huge difference.* Wonkhe. https://wonkhe.com/blogs/criminalising-essay-mills-will-make-a-huge-difference/

Draper, M., & Crossman, G. (2019). *Essay mills and the case for legislation.* https://www.qaa.ac.uk/docs/qaa/guidance/essay-mills-and-the-case-for-legislation.pdf
Draper, M. J., & Newton, P. M. (2017). A legal approach to tackling contract cheating? *International Journal for Educational Integrity, 13*(11). https://doi.org/10.1007/s40979-017-0022-5
Draper, M. J., & Reid-Hutchings, C. (2019). Are essay mills committing fraud? A further analysis of their behaviours vs the 2006 fraud act (UK). *International Journal for Educational Integrity, 15*(11). https://doi.org/10.1007/s40979-019-0050-4
Tauginienė, L., Gaižauskaitė, I., Razi, S., Glendinning, I., Sivasubramaniam, S., Marino, F., Cosentino, M., Anohina-Naumeca, A., & Kravjar, J. (2019). Enhancing the taxonomies relating to academic integrity and misconduct. *Journal of Academic Ethics, 17*, 345–361. https://doi.org/10.1007/s10805-019-09342-4
Times Higher Education (THE). (2022). Judge contract cheating law culture change, not prosecutions. *Times Higher Education.* https://www.timeshighereducation.com/news/judge-contract-cheating-law-culture-change-not-prosecutions
UK Government. (2022). Skills and post 16 education act 2022. https://www.legislation.gov.uk/en/ukpga/2022/21/contents/enacted

MIX
Papier aus verantwortungsvollen Quellen
Paper from responsible sources
FSC® C105338

If you have any concerns about our products,
you can contact us on
ProductSafety@springernature.com

In case Publisher is established outside the EU,
the EU authorized representative is:
Springer Nature Customer Service Center GmbH
Europaplatz 3, 69115 Heidelberg, Germany

Printed by Libri Plureos GmbH
in Hamburg, Germany